AF510602

Plant Genetics and Breeding

Plant Genetics and Breeding

Nina Duran

Larsen & Keller
www.larsen-keller.com

Plant Genetics and Breeding
Nina Duran
ISBN: 978-1-64172-601-6 (Hardback)

© 2022 Larsen & Keller

Larsen & Keller
Published by Larsen and Keller Education,
5 Penn Plaza,
19th Floor,
New York, NY 10001, USA

Cataloging-in-Publication Data

Plant genetics and breeding / Nina Duran.
p. cm.
Includes bibliographical references and index.
ISBN 978-1-64172-601-6
1. Plant genetics. 2. Plant breeding. 3. Plant genomes. 4. Crops--Genetics. I. Duran, Nina.
QK981 .P53 2022
581.35--dc23

This book contains information obtained from authentic and highly regarded sources. All chapters are published with permission under the Creative Commons Attribution Share Alike License or equivalent. A wide variety of references are listed. Permissions and sources are indicated; for detailed attributions, please refer to the permissions page. Reasonable efforts have been made to publish reliable data and information, but the authors, editors and publisher cannot assume any responsibility for the validity of all materials or the consequences of their use.

Trademark Notice: All trademarks used herein are the property of their respective owners. The use of any trademark in this text does not vest in the author or publisher any trademark ownership rights in such trademarks, nor does the use of such trademarks imply any affiliation with or endorsement of this book by such owners.

For more information regarding Larsen and Keller Education and its products, please visit the publisher's website www.larsen-keller.com

Table of Contents

Preface

The study of genes, genetic variation and heredity in plants is referred to as plant genetics. It is a field of biology and botany that encompasses various life sciences. Plant genetics is primarily applied to the genetic modification of plants for increasing disease resistance, nutritional value and yield. The science of changing the characteristics of plants for the purpose of increasing their nutritional value is termed as plant breeding. There are a diverse range of techniques which are used in plant breeding such as marker assisted selection, genetic modification and reverse breeding. This book is a compilation of chapters that discuss the most vital concepts in the field of plant genetics and breeding. Different approaches, evaluations, methodologies and advanced studies on this field have been included in this book. This textbook will provide comprehensive knowledge to the readers.

A foreword of all Chapters of the book is provided below:

Chapter 1 - The science of changing the traits of plants to produce preferred characteristics is known as plant breeding. Plant genetics deals with the study of genes, heredity and genetic variation in plants. This is an introductory chapter which will introduce briefly all the significant aspects of plant breeding and genetics.; **Chapter 2** - Plant reproduction is the process of producing new offspring in plants. It can be accomplished through sexual or asexual reproduction. In sexual reproduction the offspring is produced by the fusion of gametes. In asexual reproduction the offspring is produced without the fusion of gametes. This chapter discusses in detail the theories and methodologies related to plant reproduction.; **Chapter 3** - Natural selection and artificial selection are the two broad classifications of plant breeding. Natural selection method involves the natural processes and functions in nature while human beings are the selecting agents in artificial selection methods. All the diverse selection methods for plant breeding have been carefully analyzed in this chapter.; **Chapter 4** - The application of the tools of molecular biology towards plant and animal breeding is known as molecular breeding. Genetic markers such as molecular markers, morphological markers and biochemical markers are used widely in plant breeding. This chapter discusses in detail the theories and methodologies related to plant molecular breeding.; **Chapter 5** - The negative impact of non-living elements on a living organism in a particular environment is known as abiotic stress. Plants are exposed to several kinds of abiotic stress such as extreme temperatures, droughts, floods and high salinity. The chapter closely examines the key concepts of plant breeding to provide an extensive understanding of the different methods which are used to increase resistance for abiotic stress in plants; **Chapter 6** - The direct manipulation of an organism's genes using biotechnology is referred to as genetic modification, or genetic engineering. Genetically modified plants are known as transgenic plants. Plants are genetically modified for various reasons such as for ornamental use, to increase yield in crops and for conservation purposes. This chapter discusses in detail the theories and methodologies related to genetic modification in plants.

I would like to thank the entire editorial team who made sincere efforts for this book and my family who supported me in my efforts of working on this book. I take this opportunity to thank all those who have been a guiding force throughout my life.

Nina Duran

1

Plant Breeding and Genetics

The science of changing the traits of plants to produce preferred characteristics is known as plant breeding. Plant genetics deals with the study of genes, heredity and genetic variation in plants. This is an introductory chapter which will introduce briefly all the significant aspects of plant breeding and genetics.

Plant Breeding

Plant breeding is the application of genetic principles to produce plants that are more useful to humans. This is accomplished by selecting plants found to be economically or aesthetically desirable, first by controlling the mating of selected individuals, and then by selecting certain individuals among the progeny. Such processes, repeated over many generations, can change the hereditary makeup and value of a plant population far beyond the natural limits of previously existing populations.

'Double Delight,' a hybrid tea rose (Rosa 'Double Delight'). The flowers are highly fragrant and bloom continuously throughout the season.

Goals

The plant breeder usually has in mind an ideal plant that combines a maximum number of desirable characteristics. These characteristics may include resistance to diseases and insects; tolerance to heat, soil salinity, or frost; appropriate size, shape, and time to maturity; and many other general and specific traits that contribute to improved adaptation to the environment, ease in growing and handling, greater yield, and better quality. The breeder of horticultural plants must also consider aesthetic appeal. Thus the breeder can rarely focus attention on any one characteristic but must take into account the manifold traits that make the plant more useful in fulfilling the purpose for which it is grown. Plant breeding is an important tool in promoting global food security, and many staple crops have been bred to better withstand extreme weather conditions associated with global warming, such as drought or heat waves.

Increase of Yield

One of the aims of virtually every breeding project is to increase yield. This can often be brought about by selecting obvious morphological variants. One example is the selection of dwarf, early maturing varieties of rice. These dwarf varieties are sturdy and give a greater yield of grain. Furthermore, their early maturity frees the land quickly, often allowing an additional planting of rice or other crop the same year.

Another way of increasing yield is to develop varieties resistant to diseases and insects. In many cases the development of resistant varieties has been the only practical method of pest control. Perhaps the most important feature of resistant varieties is the stabilizing effect they have on production and hence on steady food supplies. Varieties tolerant to drought, heat, or cold provide the same benefit.

Modifications of Range and Constitution

Another common goal of plant breeding is to extend the area of production of a crop species. A good example is the modification of grain sorghum since its introduction to the United States in the 1750s. Of tropical origin, grain sorghum was largely confined to the southern Plains area and the Southwest, but earlier-maturing varieties were developed, and grain sorghum is now an important crop as far north as North Dakota.

Development of crop varieties suitable for mechanized agriculture has become a major goal of plant breeding in recent years. Uniformity of plant characters is very important in mechanized agriculture because field operations are much easier when the individuals of a variety are similar in time of germination, growth rate, size of fruit, and so on. Uniformity in maturity is, of course, essential when crops such as tomatoes and peas are harvested mechanically.

The nutritional quality of plants can be greatly improved by breeding. For example, it is possible to breed varieties of corn (maize) much higher in lysine than previously existing varieties. Breeding high-lysine maize varieties for those areas of the world where maize is the major source of this nutritionally essential amino acid has become a major goal in plant breeding. This "biofortification" of food crops, a term which also includes genetic modification, has been shown to improve nutrition and is especially useful in developing areas where nutritional deficiencies are common and medical infrastructure may be lacking.

In breeding ornamental plants, attention is paid to such factors as longer blooming periods, improved keeping qualities of flowers, general thriftiness, and other features that contribute to usefulness and aesthetic appeal. Novelty itself is often a virtue in ornamentals, and the spectacular, even the bizarre, is often sought.

Evaluation of Plants

The appraisal of the value of plants so that the breeder can decide which individuals should be discarded and which allowed to produce the next generation is a much more difficult task with some traits than with others.

Qualitative Characters

The easiest characters, or traits, to deal with are those involving discontinuous, or qualitative, differences that are governed by one or a few major genes. Many such inherited differences exist, and they frequently have profound effects on plant value and utilization. Examples are starchy versus sugary kernels (characteristic of field and sweet corn, respectively) and determinant versus indeterminant habit of growth in green beans (determinant varieties are adapted to mechanical harvesting). Such differences can be seen easily and evaluated quickly, and the expression of the traits remains the same regardless of the environment in which the plant grows. Traits of this type are termed highly heritable.

Quantitative Characters

In other cases, however, plant traits grade gradually from one extreme to another in a continuous series, and classification into discrete classes is not possible. Such variability is termed quantitative. Many traits of economic importance are of this type; e.g., height, cold and drought tolerance, time to maturity, and, in particular, yield. These traits are governed by many genes, each having a small effect. Although the distinction between the two types of traits is not absolute, it is nevertheless convenient to designate qualitative characters as those involving discrete differences and quantitative characters as those involving a graded series.

Quantitative characters are much more difficult for the breeder to control, for three main reasons:

1. The sheer numbers of the genes involved make hereditary change slow and difficult to assess;
2. The variations of the traits involved are generally detectable only through measurement and exacting statistical analyses; and
3. Most of the variations are due to the environment rather than to genetic endowment;

For example, the heritability of certain traits is less than 5 percent, meaning that 5 percent of the observed variation is caused by genes and 95 percent is caused by environmental influences.

It follows that carefully designed experiments are required to distinguish plants that are superior because they carry desirable genes from those that are superior because they happen to grow in a favourable site.

Methods of Plant Breeding

Mating Systems

Angiosperm mating systems devolve about the type of pollination, or transferal of pollen from flower to flower. A flower is self-pollinated (a “selfer”) if pollen is transferred to it from any flower of the same plant and cross-pollinated (an “outcrosser” or “outbreeder”) if the pollen comes from a flower on a different plant. About half of the more important cultivated plants are naturally cross-pollinated, and their reproductive systems include various devices that encourage cross-pollination—e.g., protandry (pollen shed before the ovules are mature, as in the carrot and walnut), dioecy (male and female parts are borne on different plants, as in the date palm, asparagus, and hops), and genetically determined self-incompatibility (inability of pollen to grow on the stigma of the same plant, as in white clover, cabbage, and many other species).

Other plant species, including a high proportion of the most important cultivated plants such as wheat, barley, rice, peas, beans, and tomatoes, are predominantly self-pollinating. There are relatively few reproductive mechanisms that promote self-pollination; the most positive of which is failure of the flowers to open (cleistogamy), as in certain violets. In barley, wheat, and lettuce the pollen is shed before or just as the flowers open, and in the tomato pollination follows opening of the flower, but the stamens form a cone around the stigma. In such species there is always a risk of unwanted cross-pollination.

In controlled breeding procedures it is imperative that pollen from the desired male parent, and no other pollen, reaches the stigma of the female parent. When stamens and pistils occur in the same flower, the anthers must be removed from flowers selected as females before pollen is shed. This is usually done with forceps or scissors. Protection must also be provided from "foreign" pollen. The most common method is to cover the flower with a plastic or paper bag. When the stigma of the female parent becomes receptive, pollen from the desired male parent is transferred to it, often by breaking an anther over the stigma, and the protective bag is replaced. The production of certain hybrids is, therefore, tedious and expensive because it often requires a series of delicate, exacting, and properly timed hand operations. When male and female parts occur in separate flowers, as in corn (maize), controlled breeding is easier.

Apple blossoms being hand-pollinated.

A cross-pollinated plant, which has two parents, each of which is likely to differ in many genes, produces a diverse population of plants hybrid (heterozygous) for many traits. A self-pollinated plant, which has only one parent, produces a more uniform population of plants pure breeding (homozygous) for many traits. Thus, in contrast to outbreeders, self-breeders are likely to be highly homozygous and hence true breeding for a specified trait.

Conventional Method

As all traits of a plant are controlled by genes located on chromosomes, conventional plant breeding can be considered as the manipulation of the combination of chromosomes. In general, there are three main procedures to manipulate plant chromosome combination. First, plants of a given population which show desired traits can be selected and used for further breeding and cultivation, a process called (pure line-) selection. Second, desired traits found in different plant lines can be combined together to obtain plants which exhibit both traits simultaneously, a method termed hybridization. Heterosis, a phenomenon of increased vigor, is obtained by hybridization of inbred lines. Third, polyploidy (increased number of chromosome sets) can contribute to crop improvement. Finally, new genetic variability can be introduced through spontaneous or artificially induced mutations

Selection

Selection is the most ancient and basic procedure in plant breeding. It generally involves three distinct steps. First, a large number of selections are made from the genetically variable original population. Second, progeny rows are grown from the individual plant selections for observational purposes. After obvious elimination, the selections are grown over several years to permit observations of performance under different environmental conditions for making further eliminations. Finally, the selected and inbred lines are compared to existing commercial varieties in their yielding performance and other aspects of agronomic importance.

Hybridization

The most frequently employed plant breeding technique is hybridization. The aim of hybridization is to bring together desired traits found in different plant lines into one plant line via cross- pollination. The first step is to generate homozygous inbred lines. This is normally done by using self-pollinating plants where pollen from male flowers pollinates female flowers from the same plants. Once a pure line is generated, it is outcrossed, i. e. combined with another inbred line. Then the resulting progeny is selected for combination of the desired traits. If a trait from a wild relative of a crop species, e.g. resistance against a disease, is to be brought into the genome of the crop, a large quantity of undesired traits (like low yield, bad taste, low nutritional value) are transferred to the crop as well. These unfavorable traits must be removed by time-consuming back-crossing, i. e. repeated crossing with the crop parent. There are two types of hybrid plants: interspecific and intergeneric hybrids. Beyond this biological boundary, hybridization cannot be accomplished due to sexual incompatibility, which limits the possibilities of introducing desired traits into crop plants.

Heterosis is an effect which is achieved by crossing highly inbred lines of crop plants. Inbreeding of most crops leads to a strong reduction of vigor and size in the first generations. After six or seven generations, no further reduction in vigor or size is found. When such highly inbred plants are crossed with other inbred varieties, very vigorous, large sized, large-fruited plants may result. The term "heterosis" is used to describe the phenomenon of hybrid vigor. The most notable and successful hybrid plant ever produced is the hybrid maize. By 1919, the first commercial hybrid maize was available in the United States. Two decades later, nearly all maize was hybrid, as it is today, although the farmers must buy new hybrid seed every year, because the heterosis effect is lost in the first generation after hybridization of the inbred parental lines.

Polyploidy

Most plants are diploid. Plants with three or more complete sets of chromosomes are common and are referred to as polyploids. The increase of chromosomes sets per cell can be artificially induced by applying the chemical colchicine, which leads to a doubling of the chromosome number. Generally, the main effect of polyploidy is increase in size and genetic variability. On the other hand, polyploid plants often have a lower fertility and grow more slowly.

Induced Mutation

Instead of relying only on the introduction of genetic variability from the wild species gene pool or from other cultivars, an alternative is the introduction of mutations induced by chemicals or radiation. The mutants obtained are tested and further selected for desired traits. The site of the mutation cannot be controlled when chemicals or radiation are used as agents of mutagenesis. Because the great majority of mutants carry undesirable traits, this method has not been widely used in breeding programs.

Success in using conventional plant breeding principles and agricultural techniques reached its peak when high-yielding wheat and rice lines were cultivated in the 1960s. The doubling and tripling of productivity of these important crops in Asia signaled a agricultural revolution in the developing countries. This breakthrough in food production was termed the "Green Revolution" to describe the social, economic, and nutritional impact of the new high-yielding wheat and rice strains. Norman Borlaug was awarded the Noble Price in 1970 for his contribution in breeding new high-yielding strains of cereals. However, these strains were highly dependent on fertilizers, irrigation and agrochemicals and required energy-intensive investments. This led to degradation and loss of soils and other severe environmental problems all over the world. For these reasons, the "Green Revolution" has been both praised and damned.

Biotechnological Methods

Biotechnology is the discipline which deals with the use of living organisms or their products. In this wide sense, also traditional agriculture may be seen as a form of biotechnology. The European Federation of Biotechnology defines biotechnology as "the integrated use of biochemistry, microbiology and engineering sciences in order to achieve technological (industrial) application of the capability of microorganisms, cultured tissue cells and parts thereof". In recent years, biotechnology has developed rapidly as a practical means for accelerating success in plant breeding and improving economically important crops. The most important methods used to achieve these goals are described below:

- In Vitro Cultivation of Plant Cells and Regeneration of Plants from Cultured Cells Certain isolated somatic plant cells can be cultured in vitro (in the test tube) and are capable of proliferation and organization into tissues and eventually into complete plants. The process of regenerating whole plants out of plant cells is called in vitro regeneration. The three factors affecting plant regeneration are genotype, explant source, and culture conditions, including culture medium and environment. Different mixtures of plant hormones and other compounds in varying concentrations are used to achieve regeneration of plants from cultured cells and tissues. As the plant hormonal mechanisms are not yet understood completely, the development of in vitro cultivation and regeneration systems is still largely based on empirically testing variations of the three above mentioned factors.

- In Vitro Selection and Somaclonal Variation: Plants regenerated from cell cultures may exhibit phenotypes differing from their parent plants, sometimes at quite high frequencies. If these are heritable and affecting desirable agronomic traits, such "somaclonal variation" can be incorporated into breeding programs. However, finding of specific valuable traits by this method is largely left to chance and hence inefficient. Rather than relying on this undirected process, selection in vitro targets specific traits by subjecting large populations of cultured cells to the action of a selective agent in the Petri dish. For purposes of disease resistance, this selection can be provided by pathogens, or isolated pathotoxins that are known to have a role in pathogenesis. The selection will only allow those cells to survive and proliferate that are resistant to the challenge. Selection of cells also plays an important role in genetic engineering, where special marker genes are used to select for transgenic cells.
- Somatic Hybrid Plants: Somatic hybrid plants are plants derived from the fusion of somatic cells. Cell fusion was developed after the successful culture of a large number of plant cells stripped of their cell walls. The resulting cells without walls are referred to as protoplasts. Since also protoplasts from phylogenetically unrelated species can be fused, attempts have been made to overcome sexual incompatibility using protoplast fusion. In most cases, these attempts failed because growth and division of the fused cells did not take place when only distantly related cells were fused. Successful fusions between sexually incompatible petunia species and between potatoes and tomatoes did not lead to economically interesting products, but important contributions to the understanding of cell wall regeneration and other mechanisms were achieved.

Genetic Engineering

Genetic engineering is a term used for the directed manipulation of genes, i. e. the transfer of genes between organisms or changes in the sequence of a gene. Closely related to this field are methods which use genes or specific sequences for the identification of traits and other analytical purposes. In plant breeding, the most important and already widely used method of this kind is Restriction Fragment Length Polymorphism (RFLP). The basics of this technique are described below. For those who are not familiar with the principles of genetics, the glossary at the end of this report explains the most important terms.

Restriction Fragment Length Polymorphism (RFLP)

The techniques of traditional breeding are very time-consuming. By making crosses, also a large number of undesired genes is introduced into the genome of the plant. The undesired genes have to be "sorted out" by back-crossing. The use of Restriction Fragment Length Polymorphism greatly facilitates conventional plant breeding, because one can progress through a breeding program much faster, with smaller populations and without relying entirely on testing for the desired phenotype.

RFLP makes use of restriction endonucleases. These are enzymes which recognize and cut specific nucleotide sequences in DNA. For example, the sequence GAATTC is cut by the endonuclease EcoRl. After treatment of a plant genome which restriction endonucleases, the plant DNA is cut into pieces of different length, depending on the number of recognition sites on the DNA.

These fragments can be separated according to their size by using gel electrophoresis and are made visible as bands on the gel by hybridizing the plant DNA fragments with radiolabeled or fluorescent DNA probes. As two genomes are not identical even within a given species due to mutations, the number of restriction sites and therefore the length and numbers of DNA fragments differ, resulting in a different banding pattern on the electrophoresis gel. This variability has been termed restriction fragment length polymorphism (RFLP). The closer two organisms are related, the more the pattern of bands overlap. If a restriction site lies close to or even within an important gene, the existence of a particular band correlates with the particular trait of a plant, e.g. disease resistance. By looking at the banding pattern, breeders can identify individuals which have inherited resistance genes, and resistant plants can be selected for further breeding. The use of this technique will not only accelerate progress in plant breeding considerably, but will also facilitate the identification of resistance genes, thereby opening new possibilities in plant breeding .

Gene Transfer

In conventional breeding, the pool of available genes and the traits they code for is limited due to sexual incompatibility to other lines of the crop in question and to their wild relatives. This restriction can be overcome by using the methods of genetic engineering, which in principle allow introducing valuable traits coded for by specific genes of any organism (other plants, bacteria, fungi, animals, viruses) into the genome of any plant. The first gene transfer experiments with plants took place in the early 1980s. Normally, transgenes are inserted into the nuclear genome of a plant cell. Recently it has become possible to introduce genes into the genome of chloroplasts and other plastids (small organelles of plant cells which possess a separate genome).

Transgenic plants have been obtained using Agrobacterium-mediated DNA-transfer and direct DNA-transfer, the latter including methods such as particle bombardment, electroporation and polyethylenglycol permeabilisation. The majority of plants have been transformed using Agrobacterium mediated transformation.

Agrobacterium-mediated Gene Transfer

The Agrobacterium-mediated technique involves the natural gene transfer system resident in the bacterial plant pathogens of the genus Agrobacterium. In nature, Agrobacterium tumefaciens and Agrobacterium rhizogenes are the causative agents of the crown gall and the hairy root diseases, respectively. The utility of Agrobacterium as a gene transfer system was first recognized when it was demonstrated that these plant diseases were actually produced as a result of the transfer and integration of genes from the bacteria into the genome of the plant. Both Agrobacterium species carry a large plasmid (small circular DNA molecule) called Ti in A. tumefaciens and Ri in A. rhizogenes. A segment of this plasmid, designated T-(for transfer) DNA, is transmitted by this organism into individual plant cells, usually within wounded tissue. The T-DNA segment penetrates the plant cell nucleus and integrates randomly into the genome where it is stably incorporated and inherited like any other plant gene in a predictable, dominant Mendelian fashion. Expression of the natural genes on the T-DNA results in the synthesis of gene products that direct the observed morphological changes, i. e. tumor or hairy root formation.

In genetic engineering, the tumor inducing genes within the T-DNA which cause the plant disease are removed and replaced by foreign genes. These genes are then stably integrated into the genome of the plant after infection with the altered strain of Agrobacterium, just like the natural T-DNA. Because all tumor-inducing genes are removed, the gene transfer does not induce any disease symptoms. The most important steps in Agrobacterium-mediated gene transfer are shown in figure. This reliable method of gene transfer is well suited for plants which are susceptible to infection by Agrobacterium. Unfortunately, many species, especially economically important legumes and monocotyledons such as cereals, do not respond positively to Agrobacterium-mediated transformation. For these plants, the following methods of direct DNA uptake must be applied.

Particle Bombardment

This method, also referred to as biolistic transformation (from biological ballistics) involves coating biologically active DNA onto small tungsten or gold particles (1-5 ?m in diameter) and accelerating them into plant tissue at high velocity. The particles penetrate the plant cell wall and lodge themselves within the cell where the DNA is liberated resulting in transformation of the individual plant cell in an explant. This technique is generally less efficient than Agrobacterium-mediated transformation, but has nevertheless been particularly useful in several plant species, most notably in cereal crops. The introduction of DNA into organized, morphogenic tissues such as seeds, embryos or meristems has enabled the successful transformation and regeneration of rice, wheat, soybean and maize, thus demonstrating the enormous potential of this method.

Electroporation and Direct DNA Entry into Protoplasts

Electroporation is a process whereby very short pulses of electricity are used to reversibly permeabilize lipid bilayers of plant cell membranes. The electrical discharge enables the diffusion of macromolecules such as DNA through an otherwise impermable plasma membrane. Because the plant cell wall will not allow the efficient diffusion of many transgene constructs, protoplasts (cells without cell walls) must be prepared. This requirement presents a major obstacle for many applications as protocols making possible the regeneration of protoplasts into complete plants do not exist for many species.

Gene transfer from Agrobacterium tumefaciens to a plant cell. In nature (route A), the transfer DNA (T-DNA) contains tumor inncluding genes which lead to crown gall disease. In plant genetic engineering (route B), the tumor including genes are removed and replaced by other genes, e.g. genes which confer insect resistance. In nature as well as in genetic engineering, the agrobacteria attach to a plant cell. Then, the T-DNA is cut out of a plasmid (a small circular DNA-molecule) and is transfered to the plant cell. The T-DNA migrates to the nucleus of the plant cell and becomes incorporated into a plant chromosome. In nature (route A), the tumor inducing genes cause uncontrolled (tumor-like) growth at the site of infection after expression of the T-DNA. In genetic engineering (route B, tumor inducing genes removed by other genes), transgenic cells are selected and regenerated to whole plants. All cells of the regenerated plant now contain the transgene. DNA uptake by plant protoplasts can also be stimulated by phosphate or calcium/polyethylene glycol (PEG) coprecipitation. However, these methods all suffer from the drawback that they use protoplasts as the recipient host which often cannot be regenerated into whole plants.

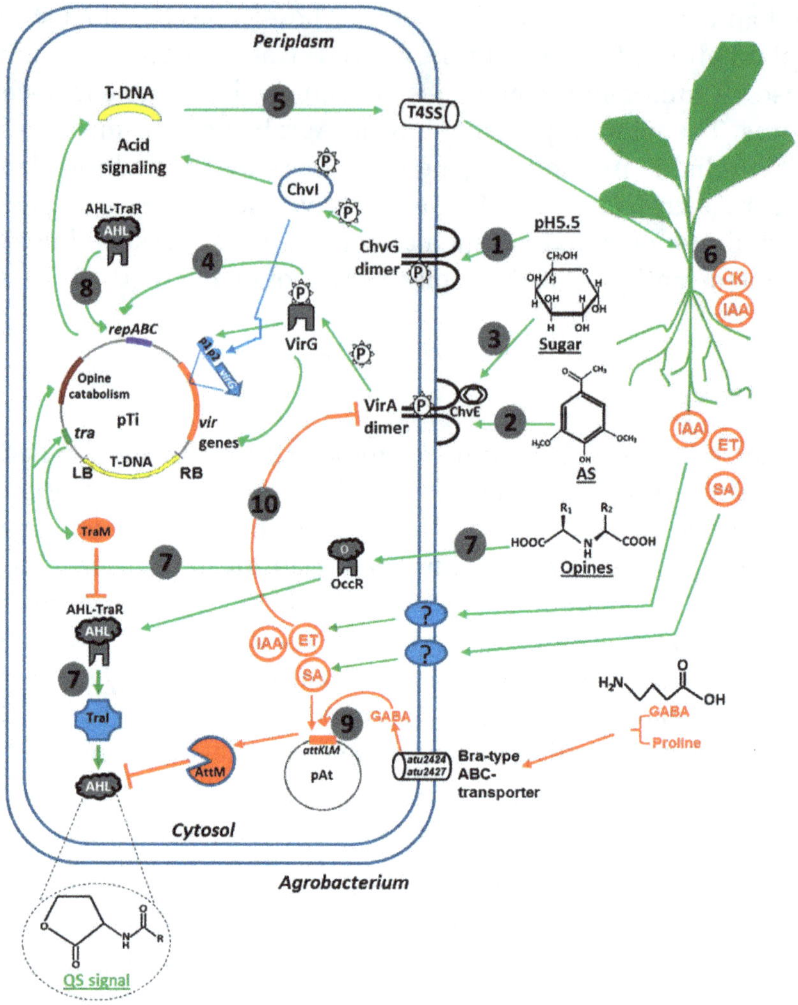

Transgene Expression

In most cases, the introduction of a gene into the plant genome will only have an effect on the plant if the transgene is expressed, i. e. transcribed into mRNA and translated into a protein. A promoter is a sequence of nucleic acids where the RNA polymerase (a complex enzyme synthesizing the mRNA transcript) attaches to the DNA template. The nature of the promoter defines (together with other expression-regulating elements), under which conditions and with which intensity a gene will be transcribed. The promoter of the 35S gene of cauliflower mosaic virus is used very frequently in plant genetic engineering. This promoter confers high-level expression of exogenous genes in most cell types from virtually all species tested. As it is often advantageous to express a transgene only in certain tissues or quantities or at certain times, a number of other promoters are available, e.g. promoters inducing gene expression after wounding or during fruit ripening only.

Methods of gene transfer currently employed result in the random integration of foreign DNA throughout the genome of recipient cells. The site of insertion may have a strong influence on the expression levels of the exogenous gene, resulting in different expression levels of an introduced gene, even if the same promoter/gene construct was used. The exact mechanism of this phenomenon are not yet fully understood.

Selection and Plant Regeneration

In a transformation experiment, the proportion of transformed cells is usually small compared to the number of cells which remain unaltered. In order to select only cells which have actually incorporated the new genes, the genes coding for the desired trait are fused to a gene which allows selection of transformed cells, so-called marker genes. The expression of the marker gene enables the transgenic cells to grow in presence of a selective agent, usually an antibiotic or a herbicide, while cells without the marker gene die. One of the most commonly used marker genes is the bacterial aminoglycoside-3' phosphotransferase gene (APH(3')II), also referred to as neomycin phosphotransferase 11 (NPTII). This gene codes for an enzyme which inactivates the antibiotics kanamycin, neomycin and G418 through phosphorylation. In addition to NPTII, a number of other antibiotic resistance genes have been used as selective markers, e.g. hygromycin phosphotransferase gene conferring resistance to hygromycin.

Another group of selective markers are herbicide tolerance genes. Herbicide tolerance has been obtained through the incorporation and expression of a gene which either detoxifies the herbicide in a similar manner as the antibiotic resistance gene products or a gene that expresses a product which acts like the herbicide target but is not affected by the herbicide. Herbicide tolerance may not only serve as a trait useful for selection in the development of transgenic plants, but also has some commercial interest. Herbicide tolerance transgenic plants are therefore among the first crops approaching market introduction.

Transformation of plant protoplasts, cells and tissues is usually only useful if the they can be regenerated into whole plants. The rates of regeneration vary greatly not only among different species, but also between cultivars of the same species. As mentioned in capter 2.2, in many cases regeneration of whole plants from cells is not possible or very difficult. Besides the ability to introduce a gene into the genome of a plant species, regeneration of intact, fertile plants out of transformed cells or tissues is the most limiting step in developing transgenic plants

Plant Genetics

A gene is a unit of heredity transferred from a parent to offspring. Regions or groups of genes make up a plant or animal's DNA structure and are what determine some characteristics of the offspring. More simply put, a gene contains a particular set of instructions.

In horticulture, a plant's genes influence the development of the plant, as they are parts of its chromosomes and are inherited through sexual propagation. Plant genes or genetics are what researchers aim to manipulate in their quest for genetically modified organisms (GMOs).

Gregor Mendel was the founding father of the genetic laws of inheritance. In his research he studied the inherited traits of pisum sativum (peas), such as their shapes, sizes, and colors. Likewise, Barbara McClintock discovered the jumping genes while she was studying maize.

Just like any other multi-cellular organism, the fundamental principles of genetics is the same in the inheritances of plants. For example, just like in animals, plants also have somatic mutations regularly. However, it is possible that the mutations can contribute to the germ line as flowers develop at the ends of branches composed of somatic cells.

These mutant branches are known as sports and can also be cultivated if they are economically desirable. Polyploidy is also more common and so plant genetics are a bit different from those of animals. (Polyploid are those having more than two paired (homologous) sets of chromosomes.

There are two ways by which genes can be transformed: the gene gun method and the agrobacterium method. The gene gun method is especially useful in transforming monocot species like corn and rice and is also known as biolistics.

The agrobacterium method has been successfully practiced in dicots, i.e. broadleaf plants, such as soybeans and tomatoes, for many years. It is also effective in monocots like grasses, including corn and rice. This method is also preferred over the gene gun method as it is easier to monitor due to a greater frequency of single-site insertions of foreign DNA

The science of genomics studies the entire DNA sequence of an organism, which covers all the chromosomes. Knowing the DNA sequence guides decisions about how to solve problems faced by that organism. In humans, for example, genomics is used to study diseases and seek cures. In plants, it is used to determine the best genes to be selected for improved crop varieties. The final goal is to predict which plants will perform best, and under which conditions. This allows plant breeders to select the plants most adapted to farmers' needs.

Traditional plant breeding can take years–or even decades, if you are breeding trees–to develop new crop varieties. Genomics allows scientists to identify which genes control important traits in a plant. They can determine which changes in the DNA sequence improve that trait. New technologies enable scientists to quickly study thousands of genes in thousands of plants. This allows plant breeders to choose, from a wide selection, the plants with the best DNA sequence at many genes and for many traits at the same time. This can often be done from a seed or seedling, instead of waiting for the plant to mature and show the desired leaf, flower, or fruit characteristics. This means a new variety can be developed in as few as two or three years–about half the time it used to take.

Plant Specific Genetics

Plants, like all other known living organisms, pass on their traits using DNA. Plants however are unique from other living organisms in the fact that they have Chloroplasts. Like mitochondria, chloroplasts have their own DNA. Like animals, plants experience somatic mutations regularly, but these mutations can contribute to the germ line with ease, since flowers develop at the ends of branches composed of somatic cells. People have known of this for centuries, and mutant branches are called "sports". If the fruit on the sport is economically desirable, a new cultivar may be obtained.

Some plant species are capable of self-fertilization, and some are nearly exclusively self-fertilizers. This means that a plant can be both mother and father to its offspring, a rare occurrence in animals. Scientists and hobbyists attempting to make crosses between different plants must take special measures to prevent the plants from self-fertilizing. In plant breeding, people create hybrids between plant species for economic and aesthetic reasons. For example, the yield of Corn has increased nearly five-fold in the past century due in part to the discovery and proliferation of hybrid corn varieties. Plant genetics can be used to predict which combination of plants may produce a plant with Hybrid vigor, or conversely many discoveries in Plant genetics have come from studying the effects of hybridization.

Plants are generally more capable of surviving, and indeed flourishing, as polyploids. Polyploid organisms have more than two sets of homologous chromosomes. For example, humans have two sets of homologous chromosomes, meaning that a typical human will have 2 copies each of 23 different chromosomes, for a total of 46. Wheat on the other hand, while having only 7 distinct chromosomes, is considered a hexaploid and has 6 copies of each chromosome, for a total of 42. In animals, inheritable germline polyploidy is less common, and spontaneous chromosome increases may not even survive past fertilization. In plants however this is no such problem, polyploid individuals are created frequently by a variety of processes, however once created usually cannot cross back to the parental type. Polyploid individuals, if capable of self-fertilizing, can give rise to a new genetically distinct lineage, which can be the start of a new species. This is often called "instant speciation". Polyploids generally have larger fruit, an economically desirable trait, and many human food crops, including wheat, maize, potatoes, peanuts, strawberries and tobacco, are either accidentally or deliberately created polyploids.

Model Organisms

Arabidopsis Thaliana

Arabidopsis thaliana, also known as thale cress, has been the model organism for the study of plant genetics. As Drosphila, a species of fruit fly, was to the understanding of early genetics, so has been arabidopsis to the understanding of plant genetics. It was the first plant to ever have its genome sequenced in the year 2000. It has a small genome, making the initial sequencing more attainable. It has a genome size of 125 Mbp that encodes about 25,000 genes. Because an incredible amount of research has been done on the plant, a database called The Arabidopsis Information Resource (TAIR) has been established as a repository for multiple data sets and information on the species. Information housed in TAIR include the complete genome sequence along with gene structure, gene product information, gene expression, DNA and seed stocks, genome maps, genetic and physical markers, publications, and information about the Arabidopsis research community. There are several ecotypes of arabidopsis that have been useful in genetic research, and the natural variation has been used to identify loci important in both biotic and abiotic stress resistance.

Arabidopsis thaliana, growing from between a crack in a sidewalk; it is considered a key model organism in plant genetics.

Brachypodium Distachyon

Brachypodium distachyon is an experimental model grass that has many attributes that make it an excellent model for temperate cereals. Unlike wheat, a tetra or hexaploid species, brachypodium is diploid with a relatively small genome (~355 Mbp) with a short life-cycle, making genomic studies on it simpler.

Nicotiana Benthamiana

Nicotiana benthamiana is often considered a model organism for both plant-pathogen and transgenic studies. Because it is easily transformed with *Agrobacterium tumefaciens*, it is used to study both the expression of pathogen genes introduced into a plant or test new genetic cassette effects.

The science of plant breeding is crucial to one of humanity's greatest challenges: the need to feed, clothe, and nourish a growing world population in the face of climate extremes, decreased water availability, demands for renewable energy, and our responsibility for environmental stewardship.

The biggest milestones and innovations in modern agrigenomics that have supported humankind in meeting this global challenge are described.

The Principles of Inheritance – 1865

Gregor Mendel, through his work on pea plants, discovered the fundamental laws of inheritance. He deduced that genes come in pairs and are inherited as distinct units, one from each parent. Mendel tracked the segregation of parental genes and their appearance in subsequent offspring as dominant or recessive traits. He recognized the mathematical patterns of inheritance from one generation to the next. Mendel's laws of heredity are:

1. The Law of Segregation: Each inherited trait is defined by a gene pair. Parental genes are randomly separated to the sex cells so that sex cells contain only one gene of the pair. Offspring, therefore, inherit one genetic allele from each parent when sex cells unite in fertilization.

2. The Law of Independent Assortment: Genes for different traits are sorted separately from one another so that the inheritance of one trait is not dependent on the inheritance of another.
3. The Law of Dominance: An organism with alternate forms of a gene will express the form that is dominant.

The genetic experiments Mendel conducted with pea plants took him eight years (1856-1863) and he published his results in 1865.

Pure Line Theory – 1903

Wilhelm Johannsen first proposed the distinction between genotype and phenotype in the study of heredity while working in Denmark in 1909. The distinction is between the hereditary dispositions of organisms, their genotypes, and the ways in which those dispositions manifest themselves in the physical characteristics of those organisms, their phenotypes. This distinction was an outgrowth of Johannsen's experiments concerning heritable variation in plants, and this influenced his pure line theory of heredity. The genotype–phenotype distinction is now considered by many to be one of the conceptual pillars of twentieth century genetics.

Hybrid Vigor – 1908

In early 1908, George Harrison Shull, then at the Cold Spring Harbor Laboratory, published a paper with the title, "the composition of a field of maize." Shull reported that inbred lines of maize showed general deterioration in yield and vigor, but that hybrids between two inbred lines immediately and completely recovered. In many cases their yield exceeded that of the varieties from which the inbred lines were derived. Furthermore, they had a highly desirable uniformity making them better suited to agriculture. In a subsequent paper in 1909, he outlined procedures leveraging the phenomenon of hybrid vigor that later became standard in corn-breeding programs.

The Double-cross Method – 1917

In 1917 Donald Forsha Jones crossed the single cross of two strains of Chester's Learning corn, with a single cross of two strains of Burr White corn. Grown in 1918 this cross, which later came to be known as a "double cross," yielded more than either of its single-cross parents and considerably more than the best open-pollinated varieties available. Within a few years corn breeding programs including the isolation of inbred strains and testing of single and double crosses had been initiated by the U.S. Department of Agriculture. By 1933 hybrid corn was in commercial production on a substantial scale and by 1949, 78 percent of the total U.S. corn acreage was planted in hybrid corn.

Transposable Elements Discovered in Maize – 1940's

Transposable elements, or transposons, are DNA sequences that can move locations within a genome, also known as "jumping genes". Discovered in corn by Nobel Prize winning geneticist Barbara McClintock in the 1940s, they were long considered by many scientists to have little role in genetics. Others however, including McClintock, hypothesized that transposons within a genome may have important roles in cells, including regulating gene expression. We now know that transposable elements are found in most organisms, making up more than 80 percent of the maize genome and nearly 50 percent of the human genome.

Agrobacterium-mediated Plant Transformation - 1977

In 1977 Marc Van Montagu and Jeff Schell discovered the gene transfer mechanism between Agrobacterium and plants, which resulted in the development of methods to alter the bacterium into an efficient delivery system for genetic engineering in plants. The plasmid transfer DNA (T-DNA), used by the bacterium to cause tumors in plants, is an ideal vehicle for genetic engineering. Directed engineering is achieved by cloning your desired gene sequence into the T-DNA that will be inserted into the host plant DNA. This process has been performed using a firefly luciferase gene to produce glowing plants. The luminescence has proven to be a useful device in the study of plant chloroplast function and as a reporter gene.

The First Biotech Plant – 1982

In 1982, the first biotech plant, an antibiotic resistant tobacco, was developed. In January 1983, at a genetic research meeting in Miami, three different teams reported success in using Agrobacterium tumefaciens, to carry new genes into plant cells, heralding the dawn of modern agricultural biotechnology.

The Gene Gun Method - 1986

A gene gun is a device for delivering exogenous DNA or transgenes to cells. The payload is an elemental particle of a heavy metal coated with DNA. This device can transform almost any type of cell, including plants, and is not limited to transformation of the nucleus; it can also transform organelles, including plastids. The original gene gun was an air pistol modified to fire dense tungsten particles. It was invented by John C. Sanford, Ed Wolf and Nelson Allen at Cornell University, and Ted Klein of DuPont, between 1983 and 1986. Their target was onions, chosen for their large cell size, and the device was used to successfully deliver particles coated with a marker gene. Genetic transformation was confirmed when the onion tissue expressed the gene. This gene gun technique, also known as biolistics, has since proven to be a versatile method of genetic modification and is generally preferred to engineer transformation-resistant crops, such as cereals.

The First Flowering Plant Genome Sequenced - 2000

The first complete genome sequence of a plant, Arabidopsis thaliana, appeared in Nature in 2000. A. thaliana is a model plant used in research to study many aspects of plant biology. Originating in Europe and Central Asia, it is a dicot, as are many important staple crops, such as the potato; commercially important food crops, such as soybean; and fiber crops, such as cotton and hardwood trees. Because germination through to senescence takes only approximately 50 days, A. thaliana offers a fast system in which to study processes that may take months or years in other flowering plants. It was also chosen as the first plant species to have its genome sequenced because of its small genome size of around 120 Mb.

The First Golden Rice Field Trial – 2004

Golden Rice is a variety which has been genetically engineered to biosynthesize beta-carotene, a precursor of vitamin A, in the edible parts of rice. It was developed with the intention of producing a fortified food to be grown and consumed in areas with a shortage of dietary vitamin A. Rice is a staple food crop for over half of the world's population, making up 30–72 percent of the energy intake for people in Asian countries, making it the perfect crop for targeting vitamin deficiencies. Golden Rice differs from its parental strain by the addition of three beta-carotene biosynthesis genes. The parental strain can naturally produce beta-carotene in its leaves, where it is involved in photosynthesis. However, the plant does not normally produce the pigment in the edible endosperm, where photosynthesis does not occur. In 2005, Golden Rice 2 was announced, which produces up to 23 times more beta-carotene than the original Golden Rice.

Crispr First Applied to Plants – 2013

In August 2013, five reports were published discussing the first application of CRISPR-Cas9 genome editing in plants. This first group of studies demonstrated the immense versatility of the technology in the field of plant biology by embracing the model species Arabidopsis thaliana and Nicotiana benthamiana as well as crops such as rice. Since then four independent groups have shown that the CRISPR-Cas9 system can introduce homozygous mutations directly into the first generation of rice and tomato transformants, highlighting the exceptionally high efficiency of the system in these species. Researchers have also demonstrated that the genetic changes induced by CRISPR are present in the germ line and segregate normally in subsequent generations without further modifications. in Arabidopsis, rice and tomato.

Inbreeding

Inbreeding is defined as the production of offspring via the mating between relatives. As the majority of plant species are hermaphroditic, they possess the capability of producing offspring from the union of gametes within one individual, i.e. selfing. Plants therefore seemingly possess the ultimate adaptation for reproductive assurance, not to mention the ability of many species to reproduce vegetatively. By selfing, plants would be putatively freed from the capricious whims of mate availability, environmental stochasticity and countless other impediments inherent to sexual reproduction. Indeed some species have given up outbreeding altogether, and they produce indistinct flowers which do not even open to allow for cross-pollination. However, the sheer variety and exuberance of flowers in the plant kingdom clearly indicates that plants as a whole have not abandoned outbreeding. Indeed, some species display mating strategies which meticulously prevent selfing and/or inbreeding. The impetus that leads to the maintenance of outbreeding is the negative consequences of inbreeding for offspring fitness, known as inbreeding depression.

Inbreeding depression in plants has provided ample material to fascinate many scientists from Darwin (1876) to modern-day molecular geneticists. The interest has lead to extensive genetic theory to explain the influence of inbreeding on offspring fitness and its subsequent impact at the population level. In many cases, the quantitative theory has synergistically arisen from experimental studies of plants in both natural and laboratory settings. Inbreeding depression effects have become especially important in the context of rare and threatened species. Through habitat destruction and other human-induced changes, many species of plants now occur in small, isolated populations. Plants in small populations are then in an 'extinction vortex', whereby they are vulnerable to the combined influences of demographic and environmental stochasiticity, as well as genetic factors. The primary genetic factor in the extinction vortex is increased inbreeding in small populations; this can lead to reduced offspring survival via inbreeding depression, exacerbating small population size. However, the importance of genetic vs. environmental effects in population extinction is the subject of many scholarly debates.

How Inbreeding occurs in Plants

Inbreeding is defined as the mating between individuals related by a shared ancestry. However, inbreeding can occur in plants via selfing, or the union of haploid gametes produced by the same genetic individual, also known as autogamy. As most flowers are hermaphroditic, selfing is a reproductive strategy present in the majority of plants. Furthermore, many plants have multiple flowers or inflorescences per individual, allowing for another type of self-fertilization, termed geitonogamy showed in early studies that this type of selfing differs from selfing within flowers in terms of the quality of offspring produced. Because plants can reproduce vegetatively and apomictically (seeds having the same genotype as their mother) discrete individuals can actually be genetic clones. Therefore, another class of selfing will be mating between genetically identical individuals. Inbreeding of course can also occur in plants through the mating between close relatives, which is termed biparental inbreeding to differentiate it from selfing. Where dispersal of propagules is limited, genetically related individuals tend to occur in patches or clumps across a landscape, increasing biparental inbreeding. Similarly, when small, less vagile insects achieve pollen dispersal, pollination between relatives within a patch will prevail.

Inbreeding Depression

With some exceptions, inbreeding reduces offspring fitness in essentially all naturally outcrossing plants and to a lesser extent in selfing species. The negative effects of mating between relatives have been noticed for many centuries. The careful breeding studies of Darwin first empirically demonstrated inbreeding depression in a wide variety of taxa. The negative effects of inbreeding have since been observed in both outcrossing and selfing species for a variety of traits with consequences for offspring fitness. Examples of traits shown to be subjects to inbreeding depression include pollen quantity, number of ovules, amount of seed, germination rate, growth rate and competitive ability. Genetic models have been developed to functionally explain reduction in fitness-related traits caused by inbreeding.

The primary genetic consequence of inbreeding is increased homozygosity. Two hypotheses for the genetic basis of inbreeding depression are put forth, both of which depend on the idea that homozygosity wll increase with inbreeding. Either the overdominance or partial dominance hypotheses are invoked to model the negative consequences of inbreeding. In the overdominance hypothesis, inbreeding depression is attributable to higher fitness of heterozygotes versus homozygotes for the loci in question. For the partial dominance hypothesis, negative fitness consequences for inbred lines are due to the fixation of recessive or partially recessive deleterious alleles. Current thought favors the latter hypothesis, where inbreeding depression is attributable to many genes of small effect. However, distinguishing between the two genetic mechanisms is complicated by linked sets of deleterious recessives that imitate overdominance effects.

The Evolution of Inbreeding in Plants

Because self-fertilization will produce the most profound inbreeding, discussion of the evolution of inbreeding in plants primarily focuses on the evolution of selfing. Altogether, plants provide an excellent system in which to study the evolution of inbreeding. Members of the same genus that are otherwise closely related often display a variety of breeding systems. The ability to self-fertilize is almost universally a derived characteristic, with self-fertilizing species arising from outcrossing species. By comparing predominantly outcrossing versus selfing species, the selective pressure surrounding the evolution of self-fertilization versus outcrossing can be estimated.

The evolution of inbreeding in plants is the product of the interplay between several selective forces, with inbreeding depression being the primary cost associated with self-fertilization. Outcrossing has inherent costs for the parent plant that will influence fitness. For example, the direct costs of copious pollen production for effective fertilization take away from the potential for seed production. The so-called 'meiotic cost' for sexual reproduction occurs because only transmitting half of the genome to the next generation lowers parental fitness. Outcrossing will also have fitness costs for offspring if it breaks up co-adapted gene complexes that confer adaptation to a local environmental conditions, i.e. outbreeding depression.

Various other ecological phenomena also foster the evolution of selfing in plants. For example, stochastic fluctuations in pollinator populations could lead to reduced seed set for insect cross-pollinated plants in years with low pollinator numbers. Populations that are subject to environmental stochasticity benefit by selfing when years are either too dry or too wet for pollen to be spread between plants. Also, founding of new populations by single individual, as is evidenced in many ruderal (weedy) species, favors self-fertilizing species. In the absence of inbreeding depression via partial dominance, selection against outcrossing would lead to the predominance of selfing. However, outcrossing is necessary to maintain genetic variability to ensure the long-term evolutionary potential of a species.

Quantitative genetic models have been developed for the evolution of inbreeding and self-fertilization in plants. Fisher first demonstrated that a gene leading to self-fertilization would rapidly spread in an outcrossing population until the entire population was selfing. However, it is unlikely that selfing is conferred by a single gene of major effect, but rather a shift to selfing is likely due to a suite of genes of small effect As demonstrated by Lande and Schemske predominant outcrossing and predominant selfing can exist as alternative stable states of a mating system. Their quantitative genetic models also confirm that highly self-fertilizing species can evolve more readily from primarily outcrossing species than vice versa . However, primarily selfing species are not necessarily an evolutionary dead end as sufficient quantitative variation can be maintained between inbred lines of a species.

How Inbreeding Depression is Measured in Plants

Breeding Studies

In selfing plant populations, the ratio of the fitness of selfed versus outcrossed progeny is used as the measure of inbreeding depression. In Darwin's classic studies of the effects of inbreeding, carefully controlled crosses were used to determine the effects of inbreeding. Controlled crosses remain to this day the primary method for measuring inbreeding depression. Generally, crossing individuals of a known or inferred relatedness produces outcrossed offspring in these types of studies. Where the fine-scale genetic structure or pedigree of a natural population is unknown, distance between parental plants is used as a proxy measure of relatedness. The inbred offspring can be similarly produced by crossing individuals within a short distance of one another, or by enclosing hermaphroditic flowers to allow them to self while preventing outcrossing. Differences in seed set between outcrossed and inbred flowers can indicate inbreeding depression if fewer seeds are produced through inbreeding. In plants without gametophytic self-incompatibility loci, reduced seed set in inbred flowers in is often attributable to lethal recessives that prevent embryo formation or other essential processes for seed set.

Measuring Inbreeding in Natural Populations

The breeding studies described above are amenable to greenhouse experiments, where manipulation of pollination can be easily achieved and controlled. However, a study by Dudash indicates that offspring fitness measured in greenhouse or garden environments may significantly underestimate inbreeding depression. More indirect methods are needed to study the effects of inbreeding depression in natural populations. Inbreeding depression has been investigated at the population level in several recent studies by examining how fitness-related traits vary between small and large populations. Reduced fitness in populations may not always be attributable to inbreeding depression due to confounding environmental variation. Experimental studies have also demonstrated density-dependent seed set known as the Allee effect to occur in small populations of plants.

Molecular Methods

Studies have made use of advances in molecular genetics for detecting inbreeding in plant populations. The advantage of this approach is that pedigree information and complex breeding experiments are not always requisite, which are often difficult to apply to natural populations. The similarity in state between two alleles at a locus determined by molecular markers can be used to estimate F, the inbreeding coefficient. Changes in F across two generations of a selfing plant can be used to estimate selection against inbred individuals, i.e. inbreeding depression. Heterozygosity at marker loci is assumed to be indicative of general levels of heterozygosity in the genome, which is expected to decrease with inbreeding. Positive correlations between fitness-related traits and heterozygosity are typically taken as evidence for inbreeding depression. The application of quantitative trait loci (QTL) methods to the issue of inbreeding depression is in its early stages of development, but has provided significant clues to its underlying genetic basis. QTLs allow for the association of observable inbreeding depression effects with discrete regions of the genome. For example in Pinus taeda, genes of major effect detected by marker linkage were primarily responsible for inbreeding depression.

Inbreeding Effects in Wild Populations

The effects of inbreeding in natural population are of interest to ecologists seeking to determine the underlying causes for observed demographic phenomena. Population level investigations of inbreeding have recently documented these effects in a number of species. Menges compared populations of an endemic prairie species Silene regia (royal catchfly) and found that seed set increased with population size. In a related study, Oostermeijer and colleagues found that offspring fitness of Gentiana pneumonanthe (marsh gentian) increased with population size. In a related rare species, Gentianella germanica (chiltern gentian), steeper population declines via reduced seed set were found in smaller populations; this reduced plant fitness occurred independently of environmental variation, indicating the inbreeding effects were the primary cause. As indicated in a study of Ipomopsis agregata (scarlet gilia), the effects of inbreeding depression may make populations more susceptible to environmental stress.

Field studies also can provide evidence to support theoretical results about the evolutionary consequences of inbreeding. The relationship between natural rates of self-fertilization and inbreeding depression can answer questions about their relative role in the evolution of breeding systems. Holtsford and Ellstrand examined how inbreeding effects vary with natural outcrossing rates in populations of Clarkia tembloriensis. The theory that inbreeding effects will be reduced in populations with high natural rates of inbreeding was not supported by their results; rather, all populations showed significant inbreeding depression for fecundity traits. In a similar study of Collinsia heterophylla, inbreeding depression effects also did not decline with as the level of natural inbreeding increased. However, these studies provide conflicting evidence for the role of purging of deleterious recessives from populations with high natural inbreeding rates.

Mechanisms to Avoid Inbreeding Depression

Plants display a number of variations in morphology and breeding systems that reduce the likelihood of inbreeding through selfing or mating with close relatives. However, avoidance of inbreeding is unlikely the only selective force leading to the evolution of these features. Many of the adaptations provide overt benefits to plant fitness beyond reducing selfing.

Varying the spatial distribution of male and female reproductive organs has evolved in many taxa of plants, and this breeding system can take on several different forms. Though most species have hermaphroditic flowers, some species have separate flowers with male and female functions. To reduce selfing, these flowers can be spatially distinct on the same individual plant (monoecy) or on separate individuals (dioecy). Many variations of these breeding systems exist, where plants may display combination of hermaphroditic and functionally male or female flowers. Plants also vary the spatial distribution of male and female organs within hermaphroditic flowers, which is known as heterostyly. This mechanism of self-pollination avoidance is present in 22 angiosperm families likely of independent evolutionary origin. In distylous species (e.g. primrose Primula vulgaris), two morphs of flowers are present: the 'pin' flowers with long styles and short stamens and the 'thrum' flowers with short styles and long stamens. The function of distyly in preventing inbreeding depends on interactions with the pollinators; pollen transfer between morphs rather than within morphs is facilitated by pollen deposition on different parts of the pollinator. Similarly, tristyly occurs when three flower morphs are present, e.g. the long-, mid- and short-styled morphs of purple loosestrife, Lythrum salicaria. Tristyly has most been studied in the context of its breakdown, where the advantage of self-fertilization outweighs the consequences of inbreeding depression.

Plants may also vary the distribution of male and female function in time, known collectively as dichogamy. Where pollen is dispersed from the anthers before the stigmas are receptive, the flowers are considered to be protandrous. Similarly, when the stigmas are receptive before pollen is shed from the anthers, the flowers are protogynous. Protandry is present in many more taxa than protogyny, indicating that dichogamy is mainly a means to control the amount of pollen exiting a flower or inflorescence.

Another means of inbreeding avoidance is via genetically mediated self-incompatibility, in which pollen with the same genotype as the maternal plant is unable to bring about fertilization. An example of this type of self-incompatibility is many apple cultivars that require other varieties to fertilize flowers for fruit to be produced. Rejection of pollen is controlled by a self-incompatibility locus known as the S-locus, where self-pollen and pollen from other plants with the same allele is rejected. Typically there are many different alleles at this locus in a population. There are two types of self-incompatibility in plants. Gametophytic incompatibility occurs where rejection is controlled by haploid genotype of the pollen itself. Sporophytic incompatibility occurs where rejection is controlled by the diploid genotype of the plant that produces the pollen.

Inbreeding and Conservation of Plants

Inbreeding is of concern to the conservation of rare plants because in many rare or threatened species high levels of inbreeding that occur in small, remnant populations. The level of inbreeding in a population is determined by a number of factors ,but is ultimately determined by population size, as this will generally determine the diversity of available mates. Furthermore, human activities like selective logging can directly increase the level of inbreeding by increasing mating between relatives in wild populations of plants. The severity of inbreeding depression is overall less in populations that have regularly inbred in the past; for example, populations of narrowly endemic species with consistent small population sizes throughout their evolutionary history. However, many species that are currently endangered or threatened now occur in small populations, which may or may not reflect historical sizes. Inbreeding depression will proximately have detrimental consequences for population growth, but long-term population survival will ultimately depend on genetic variability. Thus, inbreeding depression is an issue in the viability of populations of rare or threatened plants.

Despite the significant negative effects of inbreeding depression demonstrated in studies from Darwin to today, the relative importance of inbreeding depression as a factor in population extinction is subject to debate. There is some doubt about the importance of genetic effects in determining the survival of small populations relative to demographic and environmental stochasticity. Recent studies have provided evidence for the role of inbreeding in population extinction by demonstrating the importance of genetic diversity. Genetic 'rescue' effects have been demonstrated in a metapopulation of Silene alba (white campion), where gene influx increased the persistence of isolated populations. Though direct empirical evidence of inbreeding depression contributing to population extinction is limited, other sources of information clearly suggest that the influence of genetic factors cannot be discounted. Demographic models have predicted that inbreeding depression can significantly effect population survival, especially when it impacts the survival of adults. Circumstantial evidence from recent studies also indicates that inbreeding depression plays a role in population survival. Heschel and Paige demonstrated increased offspring fitness when small, presumedly inbred, populations were outcrossed, but did not observe such increases when large populations were similarly treated. Menges found that seed germination increased with population size in isolated populations of a threatened prairie species, Silene regia (royal catchfly). Similarly, decreased fecundity and survival in relation to genetic diversity and population size were found in populations of Gentiana pneomonanthe (marsh gentian), a rare plant of European calcareous grasslands. However, a similar study of Senecio integrifolius (groundsel) failed to detect a clear relationship between inbreeding effects and population size, indicating that other factors were likely of primary importance for this plant.

Mendel's Experiment

Mendel began a decade-long research project to investigate patterns of inheritance. Although he began his research using mice, he later switched to honeybees and plants, ultimately settling on garden peas as his primary model system. A model system is an organism that makes it easy for a researcher to investigate a particular scientific question, such as how traits are inherited. By studying a model system, researchers can learn general principles that apply to other, harder-to-study organisms or biological systems, such as humans.

Mendel studied the inheritance of seven different features in peas, including height, flower color, seed color, and seed shape. To do so, he first established pea lines with two different forms of a feature, such as tall vs. short height. He grew these lines for generations until they were pure-breeding (always produced offspring identical to the parent), then bred them to each other and observed how the traits were inherited.

In addition to recording how the plants in each generation looked, Mendel counted the exact number of plants that showed each trait. Strikingly, he found very similar patterns of inheritance for all seven features he studied:

- One form of a feature, such as tall, always concealed the other form, such as short, in the first generation after the cross. Mendel called the visible form the dominant trait and the hidden form the recessive trait.
- In the second generation, after plants were allowed to self-fertilize (pollinate themselves), the hidden form of the trait reappeared in a minority of the plants. Specifically, there were always about 3 plants that showed the dominant trait (e.g., tall) for every 1 plant that showed the recessive trait (e.g., short), making a 3:1 ratio.
- Mendel also found that the features were inherited independently: one feature, such as plant height, did not influence inheritance of other features, such as flower color or seed shape.

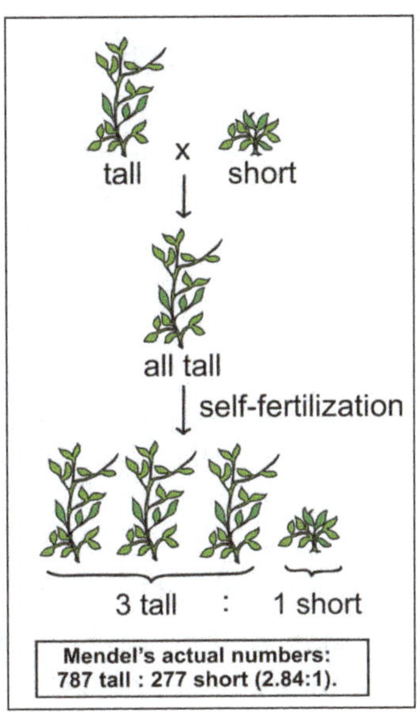

In 1865, Mendel presented the results of his experiments with nearly 30,000 pea plants to the local Natural History Society. Based on the patterns he observed, the counting data he collected, and a mathematical analysis of his results, Mendel proposed a model of inheritance in which:

- Characteristics such as flower color, plant height, and seed shape were controlled by pairs of heritable factors that came in different versions.
- One version of a factor (the dominant form) could mask the presence of another version (the recessive form).
- The two paired factors separated during gamete production, such that each gamete (sperm or egg) randomly received just one factor.
- The factors controlling different characteristics were inherited independently of one another.

Mendel's Model System: The Pea Plant

Mendel carried out his key experiments using the garden pea, Pisum sativum, as a model system. Pea plants make a convenient system for studies of inheritance, and they are still studied by some geneticists today.

Useful features of peas include their rapid life cycle and the production of lots and lots of seeds. Pea plants also typically self-fertilize, meaning that the same plant makes both the sperm and the egg that come together in fertilization. Mendel took advantage of this property to produce true-breeding pea lines: he self-fertilized and selected peas for many generations until he got lines that consistently made offspring identical to the parent (e.g., always short).

Pea plants are also easy to cross, or mate in a controlled way. This is done by transferring pollen from the anthers (male parts) of a pea plant of one variety to the carpel (female part) of a mature pea plant of a different variety. To prevent the receiving plant from self-fertilizing, Mendel painstakingly removed all of the immature anthers from the plant's flowers before the cross.

Crossing Pea Plants

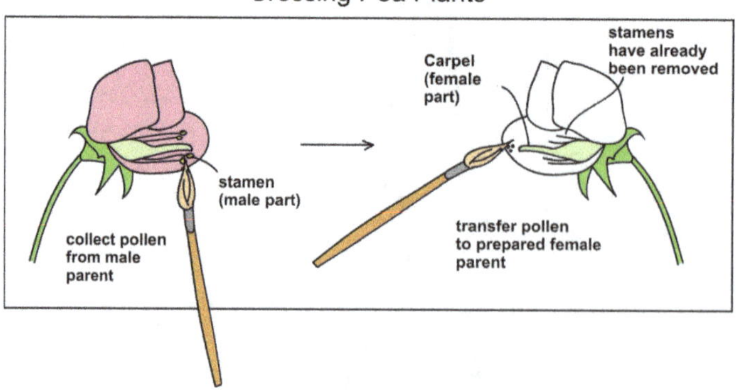

Because peas were so easy to work with and prolific in seed production, Mendel could perform many crosses and examine many individual plants, making sure that his results were consistent (not just a fluke) and accurate (based on many data points).

Mendel's Experimental Setup

Once Mendel had established true-breeding pea lines with different traits for one or more features of interest (such as tall vs. short height), he began to investigate how the traits were inherited by carrying out a series of crosses.

First, he crossed one true-breeding parent to another. The plants used in this initial cross are called the P generation, or parental generation.

Mendel collected the seeds from the P generation cross and grew them up. These offspring were called the F_1 generation, short for first filial generation.

Once Mendel examined the F_1 plants and recorded their traits, he let them self-fertilize naturally, producing lots of seeds. He then collected and grew the seeds from the F_1 plants to produce an F_2 generation, or second filial generation. Again, he carefully examined the plants and recorded their traits.

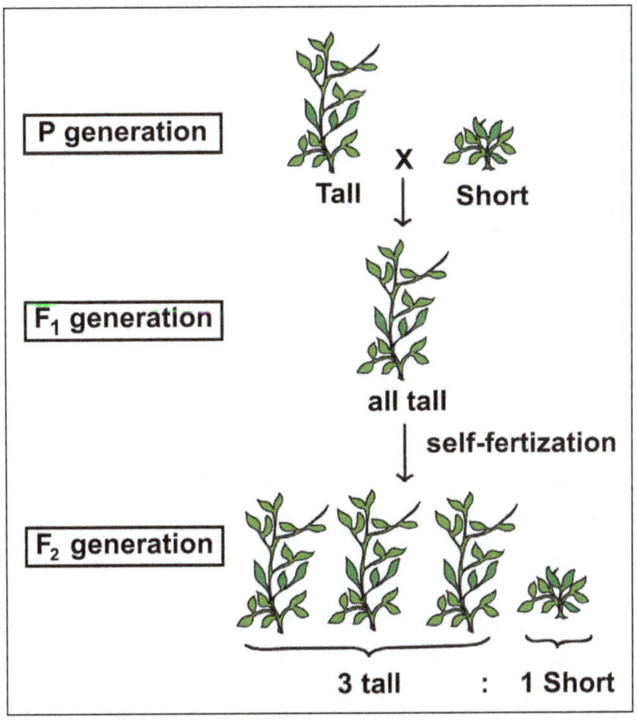

Mendel's experiments extended beyond the F_2 generation to $F_{3,}$ F_4 and later generations, but his model of inheritance was based mostly on the first three generation (P, F_1 and F_2).

Mendel didn't just record what his plants looked like in each generation (e.g., tall vs. short). Instead, he counted exactly how many plants with each trait were present.

Monohybrid Cross

It is the cross between two pea plants which have one pair of contrasting characters. For Example, a cross between a tall pea plant and a short (dwarf) plant. The following diagram explains this in detail.

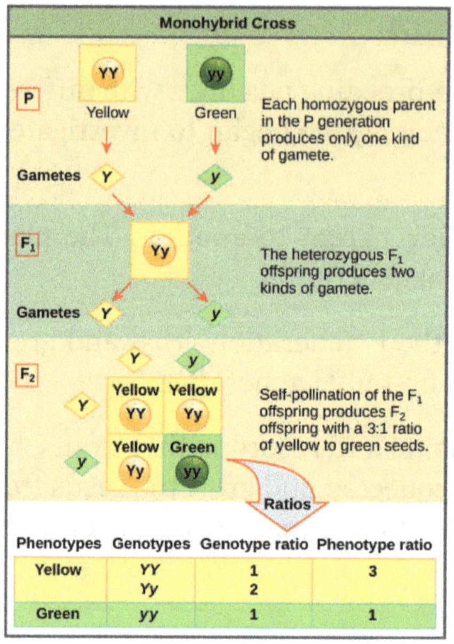

Observations

- In the first generation (F_1), the progeny were tall. There was no medium height plant.
- In the second generation (F_2), 1/4th of the offspring were short and 3/4 were tall.
- The Phenotypic ratio in F_2 is 3: 1 (3 tall: 1 short).
- The Genotypic ratio in F_2 – 1: 2: 1 – (TT: Tt: tt).
- For a plant to be tall, a single copy of "T" is enough. But if a plant has to be short, both the copies should be "t".
- Characters like 'T' are the dominant traits as they are expressed and 't' are recessive traits as they remain suppressed.

Dihybrid Cross

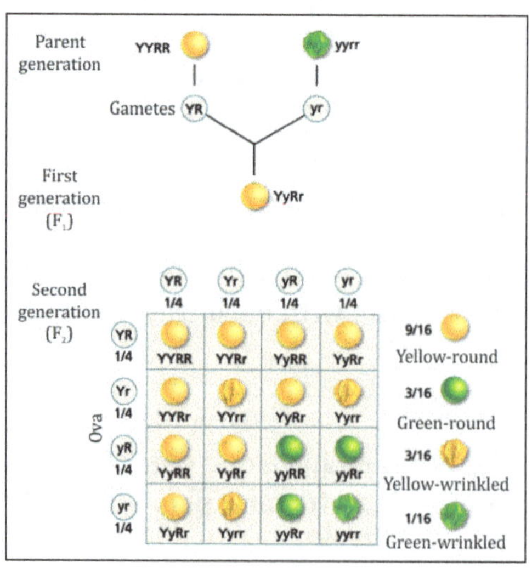

It is the cross between two plants which have two pairs of contrasting characters. This takes into consideration alternative traits of two different characters. For example, a cross between one pea plant with round and green seeds and the other pea plant having wrinkled and yellow seeds. The following diagram explains the dihybrid cross in detail.

Observations

- The F_1 generation is 100% hybrid. When RRyy crosses with rrYY, all were Rr Yy with round and yellow seeds in the first generation. The Round and Yellow seeds are the dominant characters.
- In F_2, the phenotype ratio is 9:3:3:1. The genotype ratio is a very complex one.
- This shows that the genes are inherited independently of each other.

Mendel generalized the results of his pea-plant experiments into four postulates, some of which are sometimes called "laws," that describe the basis of dominant and recessive inheritance in diploid organisms, these laws summarize the basics of classical genetics.

Pairs of Unit Factors or Genes

Mendel proposed first that paired unit factors of heredity were transmitted faithfully from generation to generation by the dissociation and reassociation of paired factors during gametogenesis and fertilization, respectively. After he crossed peas with contrasting traits and found that the recessive trait resurfaced in the F_2 generation, Mendel deduced that hereditary factors must be inherited as discrete units. This finding contradicted the belief at that time that parental traits were blended in the offspring.

Alleles can be Dominant or Recessive

Mendel's law of dominance states that in a heterozygote, one trait will conceal the presence of another trait for the same characteristic. Rather than both alleles contributing to a phenotype, the dominant allele will be expressed exclusively. The recessive allele will remain "latent" but will be transmitted to offspring by the same manner in which the dominant allele is transmitted. The recessive trait will only be expressed by offspring that have two copies of this allele and these offspring will breed true when self-crossed.

The child in the photo expresses albinism, a recessive trait.

Since Mendel's experiments with pea plants, other researchers have found that the law of dominance does not always hold true. Instead, several different patterns of inheritance have been found to exist.

Equal Segregation of Alleles

Observing that true-breeding pea plants with contrasting traits gave rise to F_1 generations that all expressed the dominant trait and F_2 generations that expressed the dominant and recessive traits in a 3:1 ratio, Mendel proposed the law of segregation. This law states that paired unit factors (genes) must segregate equally into gametes such that offspring have an equal likelihood of inheriting either factor. For the F_2 generation of a monohybrid cross, the following three possible combinations of genotypes could result: homozygous dominant, heterozygous, or homozygous recessive. Because heterozygotes could arise from two different pathways (receiving one dominant and one recessive allele from either parent), and because heterozygotes and homozygous dominant individuals are phenotypically identical, the law supports Mendel's observed 3:1 phenotypic ratio. The equal segregation of alleles is the reason we can apply the Punnett square to accurately predict the offspring of parents with known genotypes. The physical basis of Mendel's law of segregation is the first division of meiosis, in which the homologous chromosomes with their different versions of each gene are segregated into daughter nuclei. The role of the meiotic segregation of chromosomes in sexual reproduction was not understood by the scientific community during Mendel's lifetime.

Independent Assortment

Mendel's law of independent assortment states that genes do not influence each other with regard to the sorting of alleles into gametes, and every possible combination of alleles for every gene is equally likely to occur. The independent assortment of genes can be illustrated by the dihybrid cross, a cross between two true-breeding parents that express different traits for two characteristics. Consider the characteristics of seed color and seed texture for two pea plants, one that has green, wrinkled seeds (yyrr) and another that has yellow, round seeds (YYRR). Because each parent is homozygous, the law of segregation indicates that the gametes for the green/wrinkled plant all are yr, and the gametes for the yellow/round plant are all YR. Therefore, the F_1 generation of offspring all are YyRr.

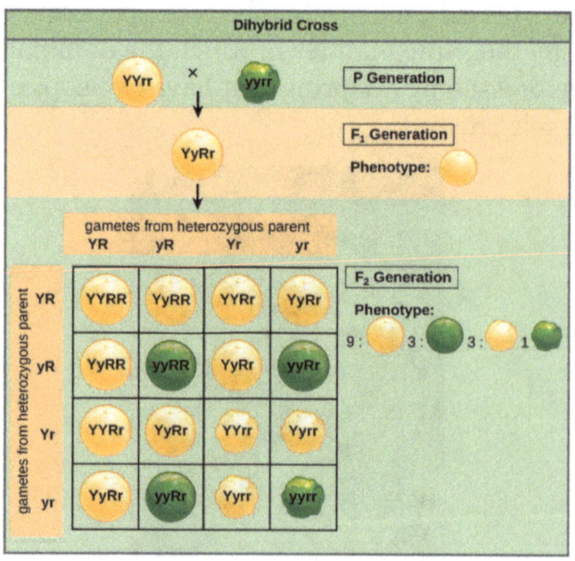

This dihybrid cross of pea plants involves the genes for seed color and texture.

For the F_2 generation, the law of segregation requires that each gamete receive either an R allele or an r allele along with either a Y allele or a y allele. The law of independent assortment states that a gamete into which an r allele sorted would be equally likely to contain either a Y allele or a y allele. Thus, there are four equally likely gametes that can be formed when the YyRr heterozygote is self-crossed, as follows: YR, Yr, yR, and yr. Arranging these gametes along the top and left of a 4 × 4 Punnett square gives us 16 equally likely genotypic combinations. From these genotypes, we infer a phenotypic ratio of 9 round/yellow:3 round/green:3 wrinkled/yellow:1 wrinkled/green. These are the offspring ratios we would expect, assuming we performed the crosses with a large enough sample size.

Because of independent assortment and dominance, the 9:3:3:1 dihybrid phenotypic ratio can be collapsed into two 3:1 ratios, characteristic of any monohybrid cross that follows a dominant and recessive pattern. Ignoring seed color and considering only seed texture in the above dihybrid cross, we would expect that three quarters of the F_2 generation offspring would be round, and one quarter would be wrinkled. Similarly, isolating only seed color, we would assume that three quarters of the F_2 offspring would be yellow and one quarter would be green. The sorting of alleles for texture and color are independent events, so we can apply the product rule. Therefore, the proportion of round and yellow F_2 offspring is expected to be (3/4) × (3/4) = 9/16, and the proportion of wrinkled and green offspring is expected to be (1/4) × (1/4) = 1/16. These proportions are identical to those obtained using a Punnett square. Round, green and wrinkled, yellow offspring can also be calculated using the product rule, as each of these genotypes includes one dominant and one recessive phenotype. Therefore, the proportion of each is calculated as (3/4) × (1/4) = 3/16.

The law of independent assortment also indicates that a cross between yellow, wrinkled (YYrr) and green, round (yyRR) parents would yield the same F_1 and F_2 offspring as in the YYRR x yyrr cross.

The physical basis for the law of independent assortment also lies in meiosis I, in which the different homologous pairs line up in random orientations. Each gamete can contain any combination of paternal and maternal chromosomes (and therefore the genes on them) because the orientation of tetrads on the metaphase plane is random.

Gene Transfer in Plants

Gene transfer is the transfer of a foreign gene to a recipient species to alter the genome of the recipient and to give a new characteristic of the foreign gene.

The gene transfer techniques in plant genetic transformation are broadly grouped into two categories:

- Vector-mediated gene transfer
- Direct or vector less DNA transfer

The salient features of the commonly used gene (DNA) transfer methods are given in table.

Gene transfer (DNA delivery) methods in plants	
Method	Salient features
• Vector—mediated gene transfer Agtobactedum (Ti plasmid)-mediated gene transfer Plant viral vectors	Very efficient, but limited to a selected group of plants Ineffective method, hence not widely used
• Direct or vectorless DNA transfer (A) Physical methods	
Electroporation	Mostly confined to protoplasts that can be regenerated to viable plants. Many cereal crops developed.
Microprojectile (particle bombardment)	Very successful method used for a wide range of plants/ tissues. Risk of gene rearrangement high.
Microinjection	Limited use since only one cell can be microinjected at a time. Technical personnel should be highly skilled.
Liposome fusion	Confined to protoplasts that can be regenerated into viable whole plants.
Silicon carbide fibres	Requires regenerable cell suspensions. The fibres, however, require careful handling.
(B) Chemical methods	
Polyethylene glycol (PEG)-mediated	Confined to protoplasts. Regeneration of fertile plants is frequently problematical.
Diethylaminoethyl (DEAE) dextran-mediated	Does not result in stable transformants.

Method I: Vector-mediated Gene Transfer

Vector-mediated gene transfer is carried out either by Agrobacterium-mediated transformation or by use of plant viruses as vectors.

Agrobacterium-mediated Gene Transfer

Agrobacterium tumefaciens is a soil-borne, Gram-negative bacterium. It is rod shaped and motile, and belongs to the bacterial family of Rhizobiaceae. A. tumefaciens is a phytopathogen, and is treated as the nature's most effective plant genetic engineer.

Some workers consider this bacterium as the natural expert of inter-kingdom gene transfer. In fact, the major credit for the development of plant transformation techniques goes to the natural unique capability of A. tumefaciens. Thus, this bacterium is the most beloved by plant biotechnologists.

There are mainly two Species of Agrobacterium:

- A. tumefaciens that induces crown gall disease.
- A. rhizogenes that induces hairy root disease.

Crown Gall Disease and Ti Plasmid

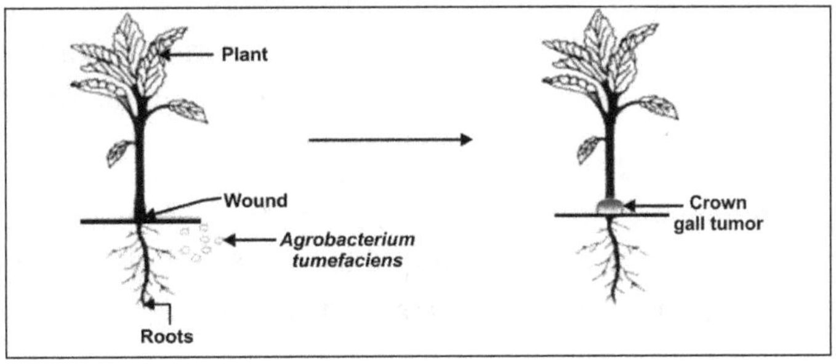

Formation of a crown gall in a plant infacted with agrobacterium tumefaciencs.

Smith and Townsend postulated that a bacterium was the causative agent of crown gall tumors, although its importance was recognized much later. As A. tumefaciens infects wounded or damaged plant tissues, in induces the formation of a plant tumor called crown gall. The entry of the bacterium into the plant tissues is facilitated by the release of certain phenolic compounds (acetosyringone, hydroxyacetosyringone) by the wounded sites.

Crown gall formation occurs when the bacterium releases its Ti plasmid (tumor- inducing plasmid) into the plant cell cytoplasm. A fragment (segment) of Ti plasmid, referred to as T-DNA, is actually transferred from the bacterium into the host where it gets integrated into the plant cell chromosome (i.e. host genome). Thus, crown gall disease is a naturally evolved genetic engineering process.

The T-DNA carries genes that code for proteins involved in the biosynthesis of growth hormones (auxin and cytokinin) and novel plant metabolites namely opines — amino acid derivatives and agropines — sugar derivatives.

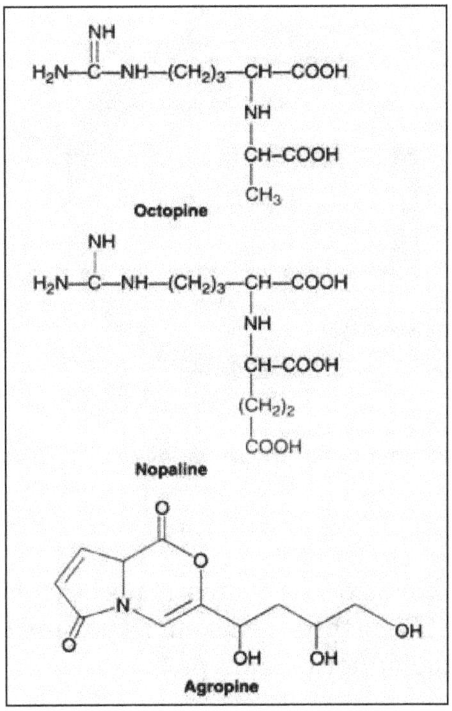

Structures of three opines – octopine nopaline and agropine

The growth hormones cause plant cells to proliferate and form the gall while opines and agropines are utilized by A. tumefaciens as sources of carbon and energy. As such, opines and agropines are not normally part of the plant metabolism (neither produced nor metabolised). Thus, A. tumefaciens genetically transforms plant cells and creates a biosynthetic machinery to produce nutrients for its own use.

As the bacteria multiply and continue infection, grown gall develops which is a visible mass of the accumulated bacteria and plant material. Crown gall formation is the consequence of the transfer, integration and expression of genes of T-DNA (or Ti plasmid) of A. tumefaciens in the infected plant.

The genetic transformation leads to the formation of crown gall tumors, which interfere with the normal growth of the plant. Several dicotyledonous plants (dicots) are affected by crown gall disease e.g. grapes, roses, stone-fruit trees.

Organization of Ti plasmid

The Ti plasmids (approximate size 200 kb each) exist as independent replicating circular DNA molecules within the Agrobacterium cells. The T-DNA (transferred DNA) is variable in length in the range of 12 to 24 kb, which depends on the bacterial strain from which Ti plasmids come. Nopaline strains of Ti plasmid have one T-DNA with length of 20 kb while octopine strains have two T-DNA regions referred to as T_L and T_R TL and TR that are respectively 14 kb and 7 kb in length.

A diagrammatic representation of a Ti plasmid is depicted in figure. The Ti plasmid has three important regions.

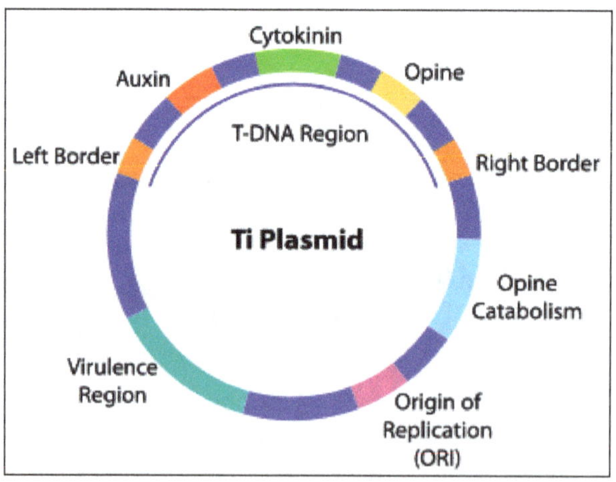

A diagrammatic representation of a Ti plasmid

T-DNA Region

This region has the genes for the biosynthesis of auxin (aux), cytokinin (cyt) and opine (ocs), and is flanked by left and right borders. These three genes-aux, cyto and ocs are referred to as oncogenes, as they are the determinants of the tumor phenotype.

T-DNA borders — A set of 24 kb sequences present on either side (right and left) of T-DNA are also transferred to the plant cells. It is now clearly established that the right border is more critical for T-DNA transfer and tumori-genesis.

Virulence Region

The genes responsible for the transfer of T-DNA into the host plant are located outside T-DNA and the region is referred to as vir or virulence region. Vir region codes for proteins involved in T-DNA transfer. At least nine vir-gene operons have been identified. These include vir A, vir G, vir B_1, vir C_1, vir D_1, D_2 and D_4, and vir E_1, and E_2.

Opine Catabolism Region

This region codes for proteins involved in the uptake and metabolisms of opines. Besides the above three, there is ori region that is responsible for the origin of DNA replication which permits the Ti plasmid to be stably maintained in A. tumefaciens.

T-DNA Transfer and Integration

The process of T-DNA transfer and it integration into the host plant genome is depicted in figure and is briefly described.

1. Signal induction to Agrobacterium: The wounded plant cells release certain chemicals-phenolic compounds and sugars which are recognized as signals by Agrobacterium. The signals induced result in a sequence of biochemical events in Agrobacterium that ultimately helps in the transfer of T-DNA of T-plasmid.
2. Attachment of Agrobacterium to plant cells: The Agrobacterium attaches to plant cells through polysaccharides, particularly cellulose fibres produced by the bacterium. Several chromosomal virulence (chv) genes responsible for the attachment of bacterial cells to plant cells have been identified.
3. Production of virulence proteins: As the signal induction occurs in the Agrobacterium cells attached to plant cells, a series of events take place that result in the production of virulence proteins. To start with, signal induction by phenolics stimulates vir A which in turn activates (by phosphorylation) vir C. This induces expression of virulence genes of Ti plasmid to produce the corresponding virulence proteins (D_1, D_2, E_2, B etc.). Certain sugars (e.g. glucose, galactose, xylose) that induce virulence genes have been identified.
4. Production of T-DNA strand: The right and left borders of T-DNA are recognized by vir D_1/vir D_2 proteins. These proteins are involved in the production single-stranded T-DNA (ss DNA), its protection and export to plant cells. The ss T-DNA gets attached to vir D_2.
5. Transfer of T-DNA out of Agrobacterium: The ss T-DNA – vir D_2 complex in association with vir G is exported from the bacterial cell. Vir B products form the transport apparatus.
6. Transfer of T-DNA into plant cells and integration: The T-DNA-vir D_2 complex crosses the plant plasma membrane. In the plant cells, T-DNA gets covered with vir E_2. This covering

protects the T-DNA from degradation by nucleases; vir D_2 and vir E_2 interact with a variety of plant proteins which influences T-DNA transport and integration.

The T-DNA-vir D_2-vir E_2 — plant protein complex enters the nucleus through nuclear pore complex. Within the nucleus, the T-DNA gets integrated into the plant chromosome through a process referred to illegitimate recombination. This is different from the homologous recombination, as it does not depend on the sequence similarity.

Hairy Root Disease of A. Rhizogenes — R_1 Plasmids

Agrobacterium rhizogenes can also infect plants. But this results in hairy roots and not crown galls as is the case with A. tumefaciens. The plasmids, of A. rhizogenes have been isolated and characterized. These plasmids, referred to as Ri plasmids, (Ri stands for Root inducing) are of different types. Some of the Ri plasmid strains possess genes that are homologous to Ti plasmid e.g. auxin biosynthetic genes.

Instead of virulence genes, Ri plasmids contain a series of open reading frames on the T-DNA. The products of these genes are involved in the metabolism of plant growth regulators which gets sensitized to auxin and leads to root formation.

Vectors of A. Rhizogenes

As it is done with A tumefaciens, vectors can be constructed by using A. rhizogenes. These vectors are alternate strategies for gene transfer. However, employment of A. rhizogene-based vectors for plant transformation is not common since more efficient systems of A. tumefaciens have been developed.

Importance of Hairy Roots

Hairy roots can be cultured in vitro, and thus are important in plant biotechnology. Hairy root systems are useful for the production of secondary metabolites, particularly pharmaceutical proteins.

Ti Plasmid-Derived Vector Systems

The ability of Ti plasmid of Agrobacterium to genetically transform plants has been described. It is possible to insert a desired DNA sequence (gene) into the T-DNA region (of Ti plasmid), and then use A. tumefaciens to deliver this gene(s) into the genome of plant cell.

In this process, Ti plasmids serve as natural vectors. However, there are several limitations to use Ti plasmids directly as cloning vectors:

- Ti plasmids are large in size (200-800 kb). Smaller vectors are preferred for recombinant experiments. For this reason, large segments of DNA of Ti plasmid, not essential for cloning, must be removed.
- Absence of unique restriction enzyme sites on Ti plasmids.
- The phytohormones (auxin, cytokinin) produced prevent the plant cells being regenerated into plants. Therefore auxin and cytokinin genes must be removed.

- Opine production in transformed plant cells lowers the plant yield. Therefore opine synthesizing genes which are of no use to plants should be removed.
- Ti plasmids cannot replicate in E. coli. This limits their utility as E. coli is widely used in recombinant experiments. An alternate arrangement is to add an origin of replication to Ti plasmid that allows the plasmid to replicate in E. coli.

Considering the above limitations, Ti plasmid- based vectors with suitable modifications have been constructed.

These vectors are mainly composed of the following components:

- The right border sequence of T-DNA which is absolutely required for T-DNA integration into plant cell DNA.
- A multiple cloning site (poly-linker DNA) that promotes the insertion of cloned gene into the region between T-DNA borders.
- An origin of DNA replication that allows the plasmids to multiply in E. coli.
- A selectable marker gene (e.g. neomycin phosphotransferase) for appropriate selection of the transformed cells.

Two types of Ti plasmid-derived vectors are used for genetic transformation of plants— co-integrate vectors and binary vectors.

Co-integrate Vector

In the co-integrate vector system, the disarmed and modified Ti plasmid combines with an intermediate cloning vector to produce a recombinant Ti plasmid.

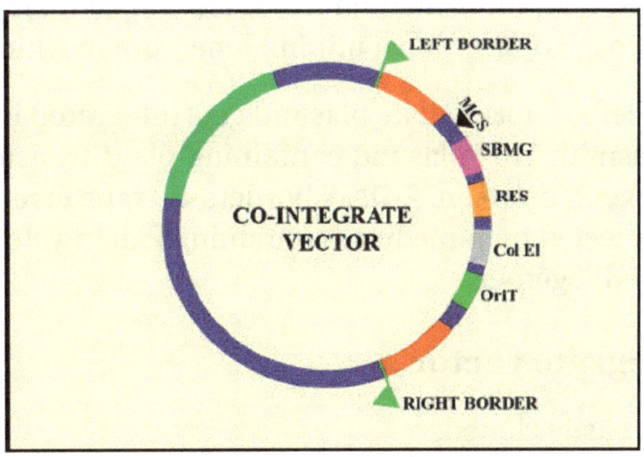

Cointegrate vector system.

Production of Disarmed Ti Plasmid

The T-DNA genes for hormone biosynthesis are removed (disarmed). In place of the deleted DNA, a bacterial plasmid (pBR322) DNA sequence is incorporated. This disarmed plasmid, also referred to as receptor plasmid, has the basic structure of T-DNA (right and left borders, virulence genes etc.) necessary to transfer the plant cells.

Construction of intermediate Vector

The intermediate vector is constructed with the following components:

- A pBR322 sequence DNA homologous to that found in the receptor Ti plasmid.
- A plant transformation marker (PTM) e.g. a gene coding for neomycin phosphotransferase II (npt II). This gene confers resistance to kanamycin in the plant cells and thus permits their isolation.
- A bacterial resistance marker e.g. a gene coding for spectinomycin resistance. This gene confers spectinomycin resistance to recipient bacterial cells and thus permits their selective isolation.
- A multiple cloning site (MCS) where foreign genes can be inserted.
- A Co/E_1 origin of replication which allows the replication of plasmid in E. coli but not in Agrobacterium.
- An oriT sequence with basis of mobilization (bom) site for the transfer of intermediate vector from E. coli to Agrobacterium.

Production and use of Co-integrate Vectors

The desired foreign gene (target-gene) is first cloned in the multiple cloning site of the intermediate vector. The cloning process is carried out in E. coli, the bacterium where the cloning is most efficient. The intermediate vector is mated with Agrobacterium so that the foreign gene is mobilised into the latter.

The transformed Agrobacterium cells with receptor Ti plasmid and intermediate vector are selectively isolated when grown on a minimal medium containing spectinomycin. The selection process becomes easy since E. coli does not grow on a minimal medium in which Agrobacterium grows.

Within the Agrobacterium cells, intermediate plasmid gets integrated into the receptor Ti plasmid to produce co-integrate plasmid. This plasmid containing plant transformation marker (e.g. npt II) gene and cloned target gene between T-DNA borders is transferred to plant cells. The transformed plant cells can be selected on a medium containing kanamycin when the plant and Agrobacterium cells are incubated together.

Advantages of Co-integrate vector

- Target genes can be easily cloned.
- The plasmid is relatively small with a number of restriction sites.
- Intermediate plasmid is conveniently cloned in E. coli and transferred to Agrobacterium.

Binary Vector

The binary vector system consists of an Agrobacterium strain along with a disarmed Ti plasmid called vir helper plasmid (the entire T-DNA region including borders deleted while vir gene is retained). It may be noted that both of them are not physically linked (or integrated). A binary vector with T-DNA can replicate in E. coli and Agrobacterium.

A diagrammatic representation of a typical binary vector system is depicted in figure. The binary vector has the following components.

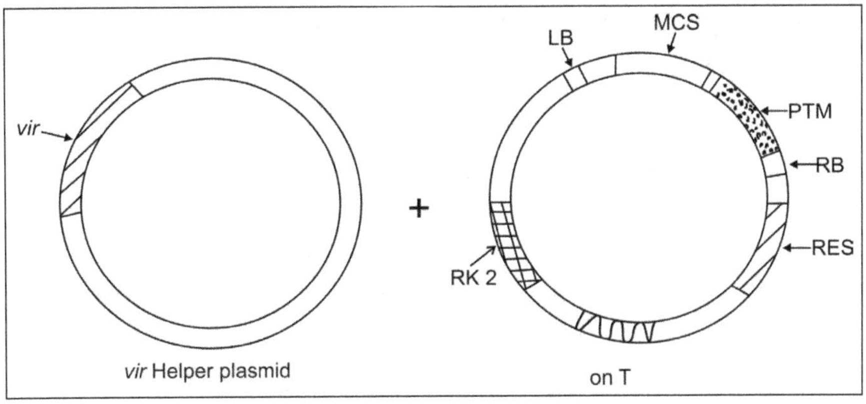

Binary vector system (vir- Ti plasmid virulence region: Lb-Left border: RB-Right border: MCS-Multiple cloning site: PTM-Plant transformation marker; RES-Bacterial resistance marker: onT-Origin of transfer site for conjugative plasmid mobilization: RK_2 -Origin of replication from plasmid).

- Left and right borders that delimit the T-DNA region.
- A plant transformation marker (PTM) e.g. npt II that confers kanamycin resistance in plant transformed cells.
- A multiple cloning site (MCS) for introducing target/foreign genes.
- A bacterial resistance marker e.g. tetracycline resistance gene for selecting binary vector colonies in E. coli and Agrobacterium.
- oriT sequence for conjugal mobilization of the binary vector from E. coli to Agrobacterium.
- A broad host-range origin of replication such as RK2 that allows the replication of binary vector in Agrobacterium.

Production and use of Binary Vector

The target (foreign) gene of interest is inserted into the multiple cloning site of the binary vector. In this way, the- target gene is placed between the right and left border repeats and cloned in E. coli. By a mating process, the binary vector is mobilised from E. coli to Agrobacterium. Now, the virulence gene proteins of T-DNA facilitate the transfer of T-DNA of the vector into plant cells.

Advantages of Binary Vectors

- The binary vector system involves only the transfer of a binary plasmid to Agrobacterium without any integration. This is in contrast to co-integrate vector system wherein the intermediate vector is transferred and integrated with disarmed Ti plasmid.
- Due to convenience, binary vectors are more frequently used than co-integrate vectors.

Plant Transformation Technique using Agrobacterium

Agrobacterium-mediated technique is the most widely used for the transformation of plants and generation of transgenic plants. The important requirements for gene transfer in higher plants through Agrobacterium mediation are listed.

- The explants of the plant must produce phenolic compounds (e.g. autosyringone) for activation of virulence genes.
- Transformed cells/tissues should be capable to regenerate into whole plants.

In general, most of the Agrobacterium-mediated plant transformations have the following basic protocol.

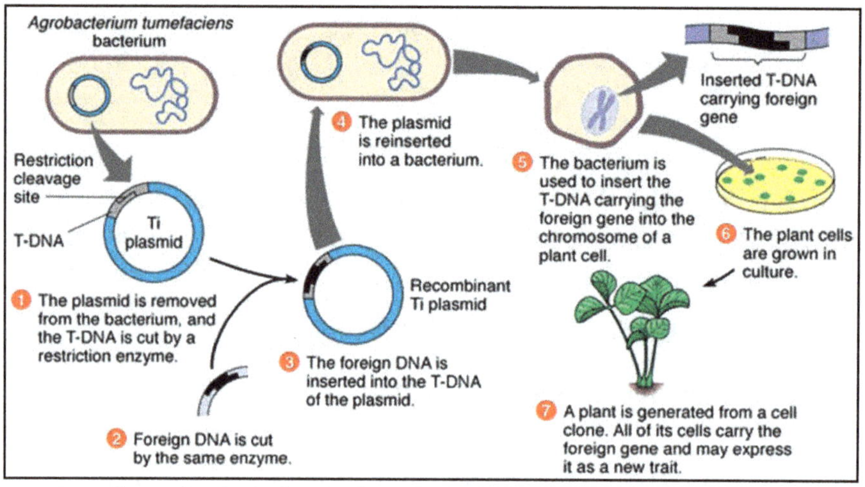

Transformation technique using Agrobadedum-medieted gene transfer.

- Development of Agrobacterium carrying the co-integrate or binary vector with the desired gene.
- Identification of a suitable explant e.g. cells, protoplasts, tissues, calluses, organs.
- Co-culture of explants with Agrobacterium.
- Killing of Agrobacterium with a suitable antibiotic without harming the plant tissue.
- Selection of transformed plant cells.
- Regeneration of whole plants.

Advantages of Agrobacterium-mediated Transformation

- This is a natural method of gene transfer.
- Agrobacterium can conveniently infect any explant (cells/tissues/organs).
- Even large fragments of DNA can be efficiently transferred.
- Stability of transferred DNA is reasonably good.
- Transformed plants can be regenerated effectively.

Limitations of Agrobacterium-mediated Transformation

- There is a limitation of host plants for Agrobacterium, since many crop plants (monocotyledons e.g. cereals) are not infected by it. In recent years, virulent strains of Agrobacterium that can infect a wide range of plants have been developed.
- The cells that regenerate more efficiently are often difficult to transform, e.g. embryonic cells lie in deep layers which are not easy targets for Agrobacterium.

Virus-Mediated Gene Transfer (Plant Viruses as Vectors)

Plant viruses are considered as efficient gene transfer agents as they can infect the intact plants and amplify the transferred genes through viral genome replication. Viruses are natural vectors for genetic engineering. They can introduce the desirable gene(s) into almost all the plant cells since the viral infections are mostly systemic.

Plant Viruses are Non-integrative Vectors

The plant viruses do not integrate into the host genome in contrast to the vectors based on T-DNA of A. tumefaciens which are integrative. The viral genomes are suitably modified by introducing desired foreign genes. These recombinant viruses are transferred, multiplied and expressed in plant cells. They spread systemically within the host plant where the new genetic material is expressed.

Criteria for a Plant Virus Vector

An ideal plant virus for its effective use in gene transfer is expected to posses the following characteristics:

- The virus must be capable of spreading from cell to cell through plasmodesmata.
- The viral genome should be able to replicate in the absence of viral coat protein and spread from cell to cell. This is desirable since the insertion of foreign DNA will make the viral genome too big to be packed.
- The recombinant viral genome must elicit little or no disease symptoms in the infected plants.
- The virus should have a broad host range.
- The virus with DNA genome is preferred since the genetic manipulations involve plant DNA.

The three groups of viruses — caulimoviruses, Gemini viruses and RNA viruses that are used as vectors for gene transfer in plants.

Caulimoviruses as Vectors

The caulimoviruses contain circular double- stranded DNA, and are spherical in shape. Caulimoviruses are widely distributed and are responsible for a number of economically important diseases in various crops. The caulimovirus group has around 15 viruses and among these cauliflower mosaic virus (CaMV) is the most important for gene transfer. The other caulimoviruses include carnation etched virus, dahlia mosaic virus, mirabilis mosaic virus and strawberry vein banding virus.

Cauliflower Mosaic Virus (CaMV)

CaMV infects many plants (e.g. members of Cruciferae, Datura) and can be easily transmitted, even mechanically. Another attractive feature of CaMV is that the infection is systemic, and large quantities of viruses are found in infected cells.

A diagrammatic view of the CaMV genetic map is depicted in figure. The genome of CaMV consists of a 8 kb (8024 bp) relaxed but tightly packed circular DNA with six major and two minor coding regions. The genes II and VII are not essential for viral infection.

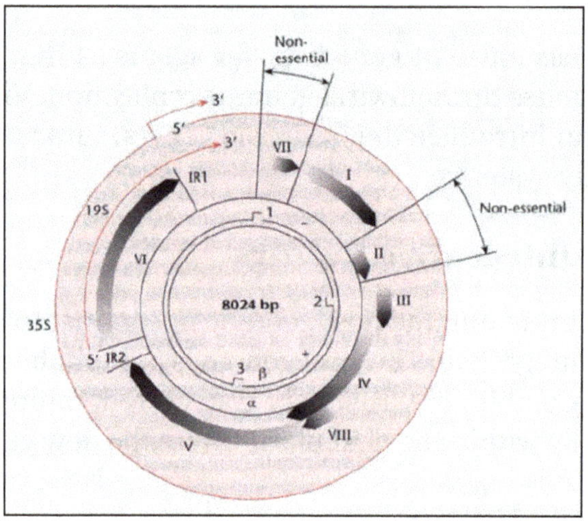

A diagrammatic representation of the genetic map of cauliflower mosaic virus genome (I... VIII represent coding regions; IR land IR 2 are intergeneric regions; The outside dotted circle represents 30S transcript The two circular lines at the centre indicate viral DNA strands).

Use of CaMV in Gene Transfer

For appropriate transmission of CaMV, the foreign DNA must be encapsulated in viral protein. Further, the newly inserted foreign DNA must not interfere with the native assembly of the virus. CaMV genome does not contain any non-coding regions wherein foreign DNA can be inserted. It is fortunate that two genes namely gene II and gene VII have no essential functions for the virus. It is therefore possible to replace one of them and insert the desired foreign gene.

Gene II of CaMV has been successfully replaced with a bacterial gene encoding dihydrofolate reductase that provides resistance to methotrexate. When the chimeric CaMV was transmitted to turnip plants, they were systemically infected and the plants developed resistance to methotrexate.

Limitations of CaMV as a Vector

- CaMV vector has a limited capacity for insertion of foreign genes.
- Infective capacity of CaMV is lost if more than a few hundred nucleotides are introduced.
- Helper viruses cannot be used since the foreign DNA gets expelled and wild-type viruses are produced.

Gemini Viruses as Vectors

The Gemini viruses are so named because they have geminate (Gemini literally means heavenly twins) morphological particles i.e. twin and paired capsid structures. These viruses are characterized by possessing one or two single-stranded circular DNAs (ss DNA). On replications, ss DNA forms an intermediate double-stranded DNA.

The Gemini viruses can infect a wide range of crop plants (monocotyledons and dicotyledons) which attract plant biotechnologists to employ these viruses for gene transfer. Curly top virus (CTV) and maize streak virus (MSV) and bean golden mosaic virus (BGMV) are among the important Gemini viruses.

It has been observed that a large number of replicative forms of a Gemini virus genome accumulate inside the nuclei of infected cells. The single-stranded genomic DNA replicates in the nucleus to form a double-stranded intermediate.

Gemini virus vectors can be used to deliver, amplify and express foreign genes in several plants/ explants (protoplasts, cultured cells). However, the serious drawback in employing Gemini viruses as vectors is that it is very difficult to introduce purified viral DNA into the plants. An alternate arrangement is to take the help of Agrobacterium and carry out gene transfer.

RNA Plant Viruses as Vectors

There are mainly two type's single-stranded RNA Viruses:

1. Mono-partite viruses: These viruses are usually large and contain undivided genomes for all the genetic information e.g. tobacco mosaic virus (TMV).
2. Multipartite viruses: The genome in these viruses is divided into small RNAs which may be in the same particle or different particles, e.g. brome mosaic virus (BMV). HMV contains four RNAs divided between three particles. Plant RNA viruses, in general, are characterized by high level of gene expression, good efficiency to infect cells and spread to different tissues. But the major limitation to use them as vectors is the difficulty of joining RNA molecules in vitro.

Use of cDNA for Gene Transfer

Complementary DNA (cDNA) copies of RNA viruses are prepared in vitro. The cDNA so generated can be used as a vector for gene transfer in plants. This approach is tedious and cumbersome. However, some success has been reported. A gene sequence encoding chloramphenicol resistance (enzyme- chloramphenicol acetyltransferase) has been inserted into brome mosaic virus genome. This gene expression, however, has been confined to protoplasts.

Limitations of Viral Vectors in Gene Transfer

The ultimate objective of gene transfer is to transmit the desired genes to subsequent generations. With virus vectors, this is not possible unless the virus is seed-transmitted. However, in case of vegetatively propagated plants, transmission of desired traits can be done e.g. potatoes. Even in these plants, there is always a risk for the transferred gene to be lost anytime. For the reasons referred above, plant biotechnologists prefer to insert the desired genes of interest into a plant chromosome.

Method II: Direct or Vector-less DNA Transfer

The term direct or vector less transfer of DNA is used when the foreign DNA is directly introduced into the plant genome. Direct DNA transfer methods rely on the delivery of naked DNA into the plant cells. This is in contrast to the Agrobacterium or vector-mediated DNA transfer which may be regarded as indirect methods. Majority of the direct DNA transfer methods are simple and effective. And in fact, several transgenic plants have been developed by this approach.

Limitations of Direct DNA transfer

The major disadvantage of direct gene transfer is that the frequency of transgene rearrangements is high. This results in higher transgene copy number, and high frequencies of gene silencing.

Types of Direct DNA transfer

The direct DNA transfer can be broadly divided into three categories.

1. Physical gene transfer methods—electro- portion, particle bombardment, microinjection, liposome fusion, silicon carbide fibres.
2. Chemical gene transfer methods—Polyethylene glycol (PEG)-mediated, diethyl amino ethyl (DEAE) dextran-mediated, calcium phosphate precipitation.
3. DNA imbibition by cells/tissues/organs.

Physical Gene Transfer Methods

An overview of the protocol for the production of transgenic plants using direct DNA delivery methods.

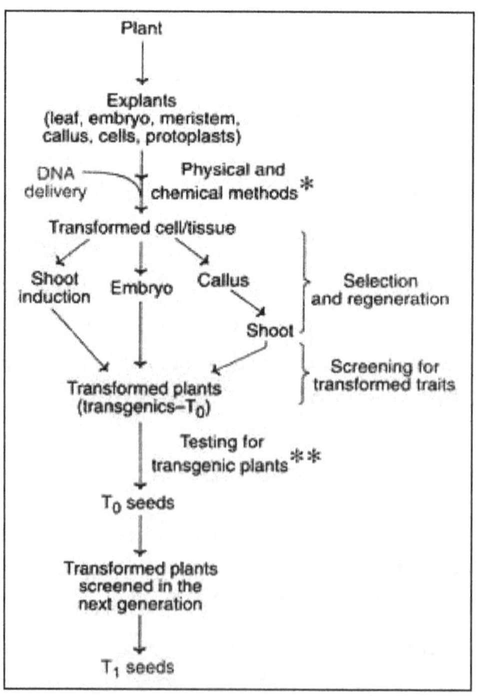

Electroporation

Electroporation basically involves the use of high field strength electrical impulses to reversibly permeabilize the cell membranes for the uptake of DNA. This technique can be used for the delivery of DNA into intact plant cells and protoplasts.

The plant material is incubated in a buffer solution containing the desired foreign/target DNA, and subjected to high voltage electrical impulses. This results in the formation of pores in the plasma membrane through which DNA enters and gets integrated into the host cell genome.

In the early years, only protoplasts were used for gene transfer by electroporation. Now a days, intact cells, callus cultures and immature embryos can be used with suitable pre- and post-electroporation treatments. Electroporation has been successfully used for the production of transgenic plants of many cereals e.g. rice, wheat, maize.

Advantages of Electroporation

- This technique is simple, convenient and rapid, besides being cost-effective.
- The transformed cells are at the same physiological state after electroporation.
- Efficiency of transformation can be improved by optimising the electrical field strength, and addition of spermidine.

Limitations of Electroporation

- Under normal conditions, the amount of DNA delivered into plant cells is very low.
- Efficiency of electroporation is highly variable depending on the plant material and the treatment conditions.
- Regeneration of plants is not very easy, particularly when protoplasts are used.

Particle Bombardment (Biolistics)

Particle (or micro projectile) bombardment is the most effective method for gene transfer, and creation of transgenic plants. This method is versatile due to the fact that it can be successfully used for the DNA transfer in mammalian cells and microorganisms.

The micro projectile bombardment method was initially named as biolistics by its inventor Sanford. Biolistics is a combination of biological and ballistics. There are other names for this technique- particle gun, gene gun, bio blaster. A diagrammatic representation of micro projectile bombardment system for the transfer of genes in plants is depicted in figure.

Micro carriers (micro projectiles), the tungsten or gold particles coated with DNA, are carried by macro carriers (macro projectiles). These macro-carriers are inserted into the apparatus and pushed downward by rupturing the disc.

The stopping plate does not permit the movement of macro carrier while the micro carriers (with DNA) are propelled at a high speed into the plant material. Here the DNA segments are released which enter the plant cells and integrate with the genome.

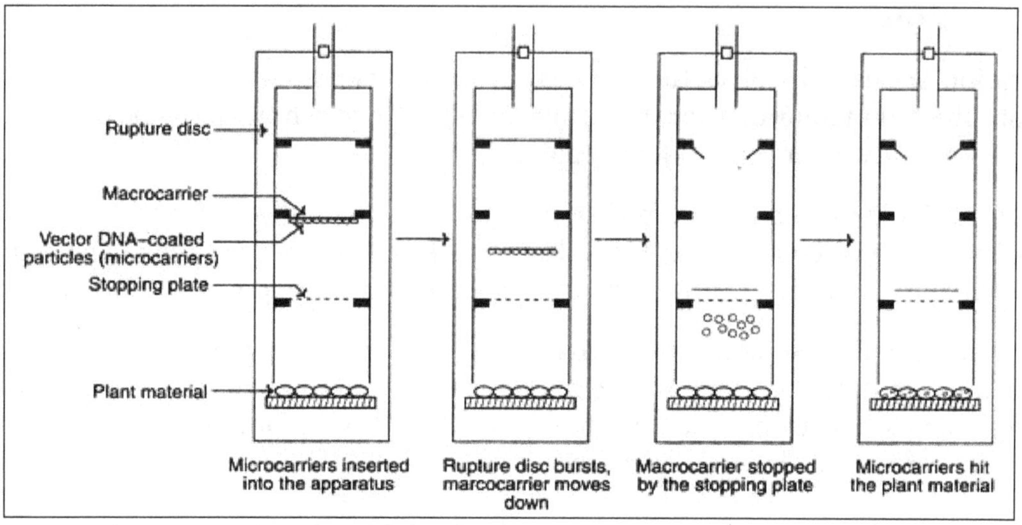

A diagrammatic representation of particle bombardment (biortstics) system for gene transfer in plants.

Plant Material used in Bombardment

Two types of plant tissue are commonly used for particle bombardment:

- Primary explants which can be subjected to bombardment that are subsequently induced to become embryo genic and regenerate.
- Proliferating embryonic tissues that can be bombarded in cultures and then allowed to proliferate and regenerate.

In order to protect plant tissues from being damaged by bombardment, cultures are maintained on high osmoticum media or subjected to limited plasmolysis.

Transgene Integration in Bombardment

It is believed (based on the gene transfer in rice by biolistics) that the gene transfer in particle bombardment is a two stage process.

- In the pre-integration phase, the vector DNA molecules are spliced together. This results in fragments carrying multiple gene copies.
- Integrative phase is characterized by the insertion of gene copies into the host plant genome.

The integrative phase facilitates further transgene integration which may occur at the same point or a point close to it. The net result is that particle bombardment is frequently associated with high copy number at a single locus. This type of single locus may be beneficial for regeneration of plants.

The Success of Bombardment

The particle bombardment technique was first introduced in 1987. It has been successfully used for the transformation of many cereals, e.g. rice, wheat, maize. In fact, the first commercial genetically modified (CM) crops such as maize containing Bt-toxin gene were developed by this approach.

A selected list of the transgenic plants (developed by biolistics) along with the sources of the plant materials used is given in table.

Plant	Cell source(s)
Rice	Embryonic callus, immature zygotic embryos
Wheat	Immature zygotic embryos
Sorghum	Immature zygotic embryos
Corn	Embryonic cell suspension, immature zygotic embryos
Barley	Cell suspension, immature zygotic embryos
Banana	Embryonic cell suspension
Sweet potato	Callus cells
Cotton	Zygotic embryos
Grape	Embryonic callus
Peas	Zygotic embryos
Peanut	Embryonic callus
Tobacco	Pollen
Alfalfa	Embryonic callus

Factors affecting Bombardment

Several attempts are made to study the various factors, and optimize the system of particle bombardment for its most efficient use. Some of the important parameters are described.

Nature of Micro Particles

Inert metals such as tungsten, gold and platinum are used as micro particles to carry DNA. These particles with relatively higher mass will have a better chance to move fast when bombarded and penetrate the tissues.

Nature of Tissues/Cells

The target cells that are capable of undergoing division are suitable for transformation. Some more details on the choice of plant material used in bombardment are already given.

Amount of DNA

The transformation may be low when too little DNA is used. On the other hand, too much DNA may result is high copy number and rearrangement of transgenes. Therefore, the quantity of DNA used should be balanced. Recently, some workers have started using the chemical aminosiloxane to coat the micro particles with low quantities of DNA adequate enough to achieve high efficiency of transformation.

Environmental Parameters

Many environmental variables are known to influence particle bombardment. These factors (temperature, humidity, photoperiod etc.) influence the physiology of the plant material, and consequently the gene transfer. It is also observed that some explants, after bombardment may require special regimes of light, humidity, temperature etc.

The technology of particle bombardment has been improved in recent years, particularly with regard to the use of equipment. A commercially produced particle bombardment apparatus namely PDS-1000/HC is widely used these days.

Advantages of Particle Bombardment

- Gene transfer can be efficiently done in organized tissues.
- Different species of plants can be used to develop transgenic plants.

Limitations of Particle Bombardment

- The major complication is the production of high transgene copy number. This may result in instability of transgene expression due to gene silencing.
- The target tissue may often get damaged due to lack of control of bombardment velocity.
- Sometimes, undesirable chimeric plants may be regenerated.

Microinjection

Microinjection is a direct physical method involving the mechanical insertion of the desirable DNA into a target cell. The target cell may be the one identified from intact cells, protoplasts, callus, embryos, meristems etc. Microinjection is used for the transfer of cellular organelles and for the manipulation of chromosomes.

The technique of microinjection involves the transfer of the gene through a micropipette (0.5-10.0 pm tip) into the cytoplasm/nucleus of a plant cell or protoplast. While the gene transfer is done, the recipient cells are kept immobilized in agarose embedding, and held by a suction holding pipette.

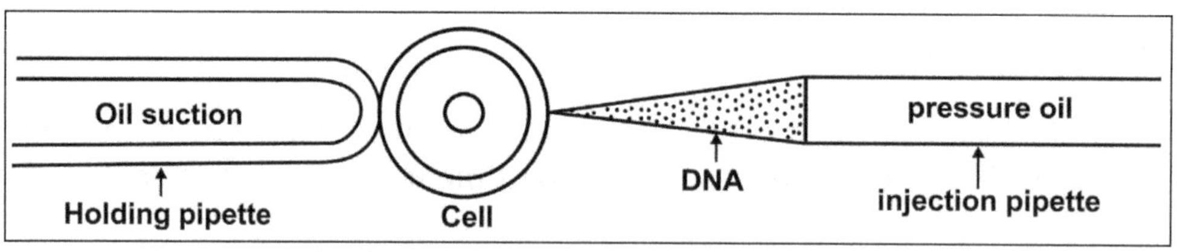

Micfoinfection of DNA by holding pipette method.

As the process of microinjection is complete, the transformed cell is cultured and grown to develop into a transgenic plant. In fact, transgenic tobacco and Brassica napus have been developed by this approach. The major limitations of microinjection are that it is slow, expensive, and has to be performed by trained and skilled personnel.

Liposome-mediated Transformation

Liposomes are artificially created lipid vesicles containing a phospholipid membrane. They are successfully used in mammalian cells for the delivery of proteins, drugs etc. Liposomes carrying genes can be employed to fuse with protoplasts and transfer the genes.

The efficiency of transformation increases when the process is carried out in conjunction with poly-ethylene glycol (PEG). Liposome-mediated transformation involves adhesion of liposomes to the protoplast surface, its fusion at the site of attachment and release of plasmids inside the cell.

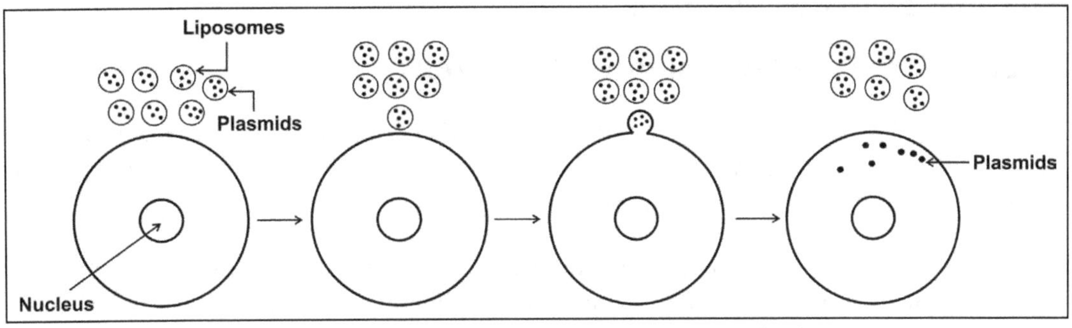

Micfoinfection of DNA by holding pipette method.

Advantages of Liposome Fusion

- Being present in an encapsulated form of liposomes, DNA is protected from environmental insults and damage.
- DNA is stable and can be stored for some time in liposomes prior to transfer.
- Applicable to a wide range of plant cells.
- There is good reproducibility in the technique.

Limitations of Liposome Fusion

The major problem with liposome-mediated transformation is the difficulty associated with the regeneration of plants from transformed protoplasts.

Silicon Carbide Fibre-mediated Transformation

The silicon carbide fibres (SCF) are about 0.3-0.6 pm in diameter and 10-100 pm in length. These fibres are capable of penetrating the cell wall and plasma membrane, and thus can deliver DNA into the cells. The DNA coated silicon carbide fibres are vortexed with 'plant material (suspension culture, calluses). During the mixing, DNA adhering to the fibres enters the cells and gets stably integrated with the host genome. The silicon carbide fibres with the trade name Whiskers are available in the market.

Advantages of SCF-mediated Transformation

- Direct delivery of DNA into intact walled cells. This avoids the protoplast isolation.
- Procedure is simple and does not involve costly equipment.

Disadvantages of SCF-mediated Transformation

- Silicon carbide fibres are carcinogenic and therefore have to be carefully handled.
- The embryonic plant cells are hard and compact and are resistant to SCF penetration.

In recent years, some improvements have been made in SCF-mediated transformation. This has helped in the transformation of rice, wheat, maize and barley by using this technique.

Chemical Gene Transfer Methods

Polyethylene Glycol (PEG)-mediated Transfer

Polyethylene glycol (PEG), in the presence of divalent cations (using Ca^{2+} , destabilizes the plasma membrane of protoplasts and renders it permeable to naked DNA. In this way, the DNA enters nucleus of the protoplasts and gets integrated with the genome.

The procedure involves the isolation of protoplasts and their suspension, addition of plasmid DNA, followed by a slow addition of 40% PEG-4000 (w/v) dissolved in mannitol and calcium nitrate solution. As this mixture is incubated, protoplasts get transformed.

Advantages of PEG-mediated Transformation

- A large number of protoplasts can be simultaneously transformed.
- This technique can be successfully used for a wide range of plant species.

Limitations of PEG-mediated Transformation

- The DNA is susceptible for degradation and rearrangement.
- Random integration of foreign DNA into genome may result in undesirable traits.
- Regeneration of plants from transformed protoplasts is a difficult task.

Deae Dextran-mediated Transfer

The desirable DNA can be complexed with a high molecular weight polymer diethyl amino ethyl (DEAE) dextran and transferred. The major limitation of this approach is that it does not yield stable trans-formants.

Calcium Phosphate Co-Precipitation-Mediated Transfer

The DNA is allowed to mix with calcium chloride solution and isotonic phosphate buffer to form DNA-calcium phosphate precipitate. When the actively dividing cells in culture are exposed to this precipitate for several hours, the cells get transformed. The success of this method is dependent on the high concentration of DNA and the protection of the complex precipitate. Addition of dimethyl sulfoxide (DMSO) increases the efficiency of transformation.

DNA Imbibition by Cells/Tissues

Some workers have seriously tried to transform cells by incubating cell suspensions, tissues, embryos and even seeds with DNA. The belief is that the DNA gets imbibed, and the cells get transformed. DNA imbibition approach has met with little or no success.

References

- Plant-breeding, science: britannica.com, Retrieved 12 July, 2019
- Plant-genetics: maximumyield.com, Retrieved 11 January, 2019
- Crop-genetics, crop-science: crops.org, Retrieved 1 August, 2019
- Griffiths, Anthony J. F.; Miller, Jeffrey H.; Suzuki, David T.; Lewontin, Richard C.; Gelbart, eds. (2000). "Genetics and the Organism: Introduction". An Introduction to Genetic Analysis (7th ed.). New York: W. H. Freeman. ISBN 978-0-7167-3520-5
- Milestones-in-plant-genetics, genomics: technologynetworks.com, Retrieved 5 March, 2019
- Inbreeding-Plants: as.wvu.edu, Retrieved 8 June, 2019
- Mendel-and-his-peas, introduction-to-heredity, classical-genetics, high-school-biology, science khanacademy.org, Retrieved 3 May, 2019
- Heredity, heredity-and-evolution, biology: toppr.com, Retrieved 5 February, 2019
- Laws-of-inheritance, biology: lumenlearning.com, Retrieved 30 April, 2019
- Dobzhansky, Theodosius; Dobzhansky, Theodosius Grigorievich (1970). Genetics of the Evolutionary Process. Columbia University Press. ISBN 9780231083065
- Methods-of-gene-transfer-in-plants, engineering, genetics: biologydiscussion.com, Retrieved 15 july, 2019

2

Plant Reproduction: Processes and Techniques

Plant reproduction is the process of producing new offspring in plants. It can be accomplished through sexual or asexual reproduction. In sexual reproduction the offspring is produced by the fusion of gametes. In asexual reproduction the offspring is produced without the fusion of gametes. This chapter discusses in detail the theories and methodologies related to plant reproduction.

Plant reproduction is the process by which plants generate new individuals, or offspring. Reproduction is either sexual or asexual. Sexual reproduction is the formation of offspring by the fusion of gametes. Asexual reproduction is the formation of offspring without the fusion of gametes. Sexual reproduction results in offspring genetically different from the parents. Asexual offspring are genetically identical except for mutation. In higher plants, offspring are packaged in a protective seed, which can be long lived and can disperse the offspring some distance from the parents. In flowering plants (angiosperms), the seed itself is contained inside a fruit, which may protect the developing seeds and aid in their dispersal.

Asexual Reproduction

The mode of reproduction by which new plants are produced without the reproductive unit of plants – flowers. This mode of reproduction occurs without the fusion of male and female gametes. Asexual reproduction produces new plants, which are the copies of the mother plant.

Asexual reproduction occurs in different kinds which includes the budding, fragmentation, vegetative propagation, and spore formation.

Types of Asexual Reproduction

Budding

Budding is the mode of asexual reproduction, wherein a new plant is developed from an outgrowth plant, called a bud. A bud is generally formed due to cell division at one particular site.

For example, if you keep a potato for a long time, you can notice a number of small growths, which are commonly referred to as 'eyes'. Each of them can be planted which will grow up like a clone of an original potato plant.

Vegetative Propagation

It is any form of asexual reproduction occurring in plants, in which new plants are produced from the vegetative parts of the plants, i.e. roots, stems or buds. Vegetative propagation in plants can occur both by naturally or also can be artificially induced by horticulturists.

The most common techniques of vegetative propagation are:

- Stems: Runners are the stems which usually grow in a horizontal form above the ground. They have the nodes where the buds are formed. These buds usually grow into a new plant.
- Roots: A new plant is developed from around, inflamed, modified roots called tubers. Example: Sweet Potato.
- Leaves: In some plants, detached leaves from the parent plant can be used to grow a new plant. They exhibit growth of small plants, called plantlets, on the edge of their leaves. Example: Bryophyllum.

Fragmentation

This is a mode of asexual reproduction in which a new plant is produced from a portion of the parent plant. Each section or a part of the plants develop into a mature, fully grown individual. Some plants possess specialized structures for reproduction through fragmentation. This type of reproduction happens naturally where the small part of the plant fall off onto soil and then begin to grow up into a new plant. This mode is often used by nurseries and greenhouses to produce plants quickly.

Spore Formation

Many plants and algae form spores in their life cycle. A spore is an asexual reproductive body, surrounded by a hard protective cover to withstand unfavorable conditions such as high temperature and low humidity. Under favorable conditions, the spores germinate and grow into new plants. Plants like moss and ferns use this mode of reproduction.

Sexual Reproduction

The type of reproduction in which the fusion of male and female gamete occurs is known as sexual reproduction. Flowers produce male and female gametes in the reproductive organs.

Process of Sexual Reproduction

This fundamental process of reproduction is DNA replication. In other words, if an organism wants to continue its species, then they need to transfer their traits to the next generation by DNA (genetic information) copying which occurs during the process of reproduction.

We know asexually reproducing organisms copy their DNA and divide themselves into new cells which are their clones. Since the process of replication is not completely reliable, there are chances of variation, but they are quite limited. Organisms need to adapt themselves to the changing environment; the creation of variants needs to be hastened for this reason. This will only happen when there is a union of two different DNAs. This highlights the significance of sexual reproduction in organisms.

If two reproducing cells simply join, the chromosome number in the new organism will be doubled. But this cannot happen because the number of chromosomes in each cell of every organism is fixed – the same holds good for the reproducing cell. Thus, the reproducing cells from both parents reduce their chromosome number into the half before fusion and are called gametes (germ cells). In simple organisms, the size and shape of gametes are almost the same. But in higher organisms, a male gamete is small and mobile while female gamete is large, immobile and they stores food for the fetus. The gametes are produced in a specialized system called the reproductive system.

Features of Sexual Reproduction

- Two parents are involved (both male and a female).
- Gamete formation and fertilization take place.
- The whole process is slow and lengthy.
- The variation occurs; offspring are different from parents, genetically and physically.

Reproductive Part of a Plant

Flowers are the reproductive parts of a plant. Male and female gametes are produced in the flower.

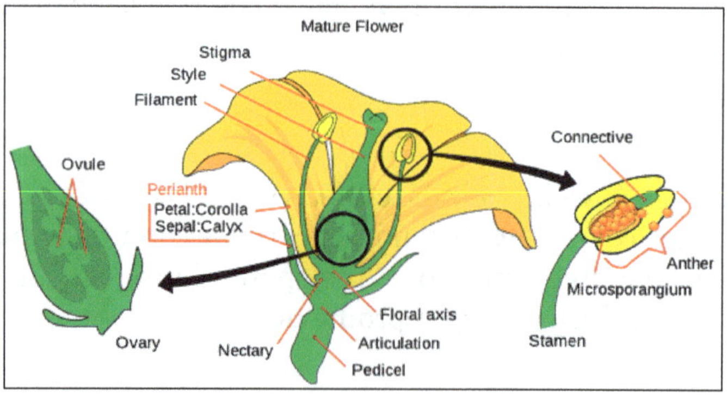

The flower consists of four main parts:

- Sepals: They are the green leaf-like structures which protect the flower in the bud stage.
- Petals: They are the coloured structures. They are the most beautiful part of the flower that attracts the birds and insects for pollination.
- Stamens: They are the male reproductive parts. Stamens consist of two parts – anther and filament. Anther is a swollen part of the stamen. The filament is a long, slender stalk which attaches the stamen to the flower. Pollen grains are formed inside the anther which produces the male gametes.

- Pistil or Carpel: It is the female reproductive part of the flower. It consists of three parts – stigma, style and the ovary. The stigma is at the top of the pistil. It receives the pollen grains during pollination. The style is a long, tube-like structure which passes the pollen grains to the ovary. The ovary is the swollen part present at the basal part of the pistil. The ovary consists of the ovule which bears the female gamete.

Uni-sexual and Bisexual Flowers

Flowers can be of two types depending upon whether both the sexual parts – male and female are present in the same flower or different flower.

- Bisexual flowers: Flowers which contain both the stamens and the pistil are known as bi-sexual flowers. For example- Rose, Hibiscus, Gulmohar, Mustard etc.
- Uni-sexual flowers: The flowers which contain either the stamens or the pistil are known as unisexual flowers. For example- Maize, Papaya, Cucumber etc.

Comparison of Asexual and Sexual Reproduction

Asexual reproduction	Sexual reproduction
In asexual reproduction, only one parent is necessary.	In sexual reproduction, two parents are necessary.
The process of gamete formation and fertilization does not occur.	The process of gamete formation and fertilization occurs.
Offsprings are exact copies of their parents.	Offsprings are not the exact copies of parents.
It generally occurs in lower organisms.	It generally occurs in higher organisms.

Pollination

Pollination is the transfer of pollen grains from the anther of one flower to the stigma of the same or another flower. It is said to be the first process of sexual fertilization in flowering plants. Pollen grains contain the male gamete and are present in the anthers of the flower.

Types of Pollination

Pollination can be of two types:

- Self-Pollination
- Cross-Pollination

Self-Pollination

When the pollen is transferred from the anthers of a flower to the stigma of the same flower, it is called as self-pollination. This form of pollination is common in hermaphrodite or dioecious plants which contain both male and female sexual parts on the same flower.

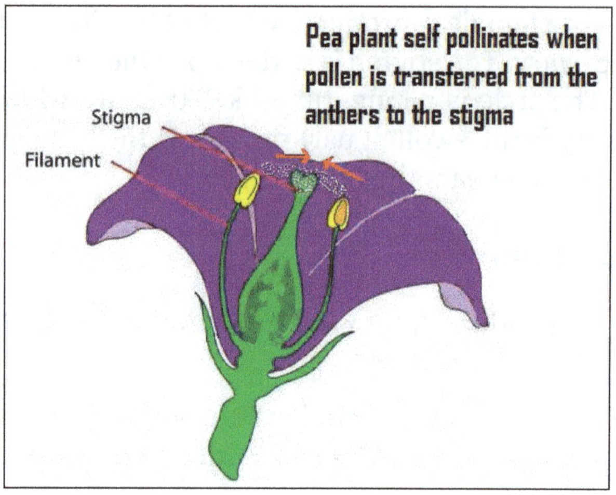

In self-pollinating plants, there is less dependence on the external factors to cause pollination. These plants depend on wind or other smaller insects that visit the flower regularly. In self- pollinating flowers, the anthers, and stigma are of similar lengths to facilitate the transfer of pollen. Self -pollination can be further divided into two types:

- Autogamy: In this type of self-pollination, the pollen is transferred from the anthers of one flower to the stigma of the same flower.
- Geitonogamy: In this type of self-pollination, the anthers are transferred from the anthers of one flower to the stigma of another flower but on the same plant.

Advantages of Self-Pollination

- In self-pollination, there is no diversity in the genes and therefore the purity of the race is maintained.
- The plants do not depend on external factors for pollination and even smaller quantities of pollen grains produce have a good success rate in getting pollinated.
- Self-pollination ensures that recessive characters are eliminated.

Disadvantages of Self-Pollination

- Since there is no mixing up of genes, there are no new characters or features that are introduced into the lineage of the offsprings.
- Self-pollination is said to reduce the vigor and vitality of the race as there are no new features introduced.
- Without new characters introduced, the resultant offsprings' immunity to diseases reduces.

Cross-Pollination

In this type of pollination, the pollen is transferred from the anthers of one flower to the stigma of another flower. In this case, the two flowers are genetically different from each other. Cross-pollination is always dependant on another agent to cause the transfer of pollen. The agents of pollination include birds, animals, water, wind, and insects. Based on the agent of pollination, cross-pollination can be of different types:

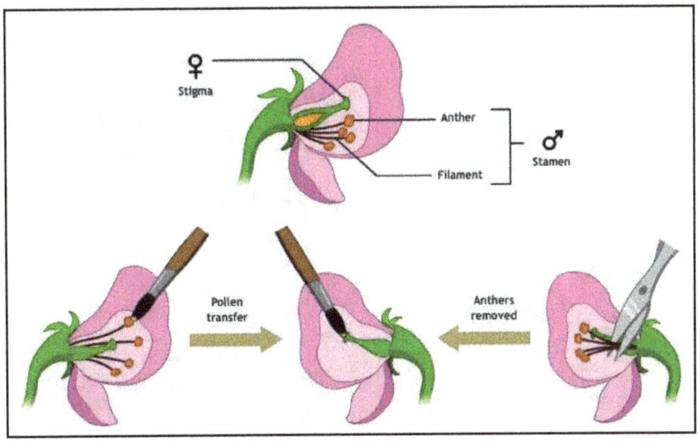

- Hydrophilous Flowers: These flowers are pollinated by water means. The flowers are often very small and inconspicuous to other agents. They do not have any fragrance or too much color on their petals. The pollen is adapted to be able to float in water.

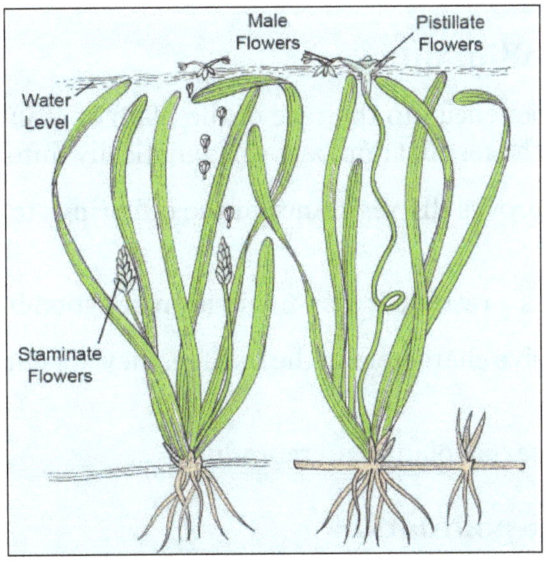

- Zoophilous flowers: In this type of pollination, the pollinating agents are animals like human beings, bats, birds etc. The zoophilous flowers have pollen that is designed to stick on to the body of the animal so that they can be easily carried from one flower to another.
- Anemophilous flowers: These flowers are pollinated by the agency of wind. These flowers, like zoophilous flowers, are small and inconspicuous. Another important feature of flowers that are wind pollinated is that they are very light so that they are easily carried by the wind. The pollen grains are very light, non-sticky and sometimes winged.
- Entomophilic flowers: These flowers are pollinated by insects. These flowers are often attractive to look at with bright petals and are fragrant to attract the insect visitors to them. They often have broad stigmas or anthers to allow the insect to perch on it. Many of the insect-pollinated flowers also secrete nectar which attracts bees, butterflies or other similar

insects to the flowers. The pollen grains in these flowers are often spiny or have extensions that help them to stick on to the body of the insects.

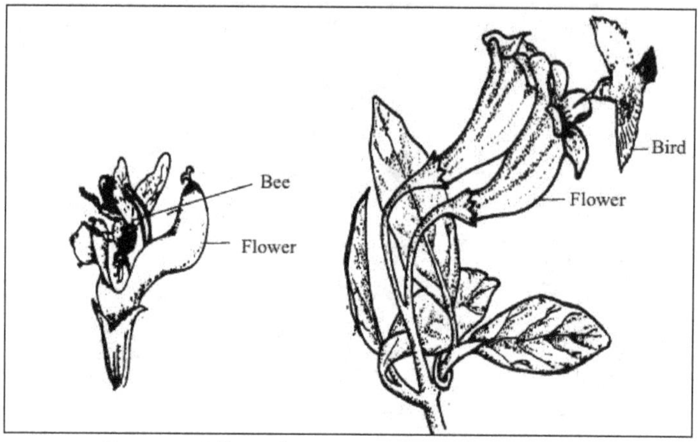

Entomophilous flower ornithophilous flower.

- Ornithophilous flowers: These flowers are pollinated by birds. Very few flowers and birds show this form of pollination.

Advantages of Cross-pollination

- Cross-pollination is beneficial to the race of the plant as it introduces new genes into the lineage as a result of the fertilization between genetically different gametes.
- Cross-pollination improves the resistance of the offsprings to diseases and changes in the environment.
- The seeds produced as a result of cross-pollination are good in vigor and vitality.
- If there are any recessive characters in the lineage, they are eliminated as a result of genetic recombination.
- It is the only way unisexual plants can reproduce.

Disadvantages of Cross-pollination

- There is a high wastage of pollen grains that need to be produced to ensure fertilization occurs.
- There are high chances that the good qualities may get eliminated and unwanted characteristics may get added due to recombination of the genes.

Methods of Pollination

Pollination by Insects

Bees are perhaps the most important pollinator of many garden plants and most commercial fruit trees. The most common species of bees are bumblebees and honeybees. Since bees cannot see the color red, bee-pollinated flowers usually have shades of blue, yellow, or other colors. Bees collect energy-rich pollen or nectar for their survival and energy needs. They visit flowers that are open during the day, are brightly colored, have a strong aroma or scent, and have a tubular shape, typically with the presence of a nectar guide. A nectar guide includes regions on the flower petals that are visible only to bees, and not to humans; it helps to guide bees to the center of the flower, thus making the pollination process more efficient. The pollen sticks to the bees' fuzzy hair, and when the bee visits another flower, some of the pollen is transferred to the second flower. Recently, there have been many reports about the declining population of honeybees. Many flowers will remain unpollinated and not bear seed if honeybees disappear. The impact on commercial fruit growers could be devastating.

Many flies are attracted to flowers that have a decaying smell or an odor of rotting flesh. These flowers, which produce nectar, usually have dull colors, such as brown or purple. They are found on the corpse flower or voodoo lily (Amorphophallus), dragon arum (Dracunculus), and carrion flower (Stapleia, Rafflesia). The nectar provides energy, whereas the pollen provides protein. Wasps are also important insect pollinators, and pollinate many species of figs.

Butterflies, such as the monarch, pollinate many garden flowers and wildflowers, which usually occur in clusters. These flowers are brightly colored, have a strong fragrance, are open during the day, and have nectar guides to make access to nectar easier. The pollen is picked up and carried on the butterfly's limbs. Moths, on the other hand, pollinate flowers during the late afternoon and night. The flowers pollinated by moths are pale or white and are flat, enabling the moths to land. One well-studied example of a moth-pollinated plant is the yucca plant, which is pollinated by the yucca moth. The shape of the flower and moth have adapted in such a way as to allow successful pollination. The moth deposits pollen on the sticky stigma for fertilization to occur later. The female moth also deposits eggs into the ovary. As the eggs develop into larvae, they obtain food from the flower and developing seeds. Thus, both the insect and flower benefit from each other in this symbiotic relationship. The corn earworm moth and Gaura plant have a similar relationship.

Pollination by Bats

In the tropics and deserts, bats are often the pollinators of nocturnal flowers such as agave, guava, and morning glory. The flowers are usually large and white or pale-colored; thus, they can be distinguished from the dark surroundings at night. The flowers have a strong, fruity, or musky fragrance and produce large amounts of nectar. They are naturally large and wide-mouthed to accommodate the head of the bat. As the bats seek the nectar, their faces and heads become covered with pollen, which is then transferred to the next flower.

Pollination by Birds

Many species of small birds, such as the hummingbird and sun birds, are pollinators for plants such as orchids and other wildflowers. Flowers visited by birds are usually sturdy and are oriented in such a way as to allow the birds to stay near the flower without getting their wings entangled in the nearby flowers. The flower typically has a curved, tubular shape, which allows access for the bird's beak. Brightly colored, odorless flowers that are open during the day are pollinated by birds. As a bird seeks energy-rich nectar, pollen is deposited on the bird's head and neck and is then transferred to the next flower it visits. Botanists have been known to determine the range of extinct plants by collecting and identifying pollen from 200-year-old bird specimens from the same site.

Pollination by Wind

Most species of conifers, and many angiosperms, such as grasses, maples and oaks, are pollinated by wind. Pine cones are brown and unscented, while the flowers of wind-pollinated angiosperm species are usually green, small, may have small or no petals, and produce large amounts of pollen. Unlike the typical insect-pollinated flowers, flowers adapted to pollination by wind do not produce nectar or scent. In wind-pollinated species, the microsporangia hang out of the flower, and, as the wind blows, the lightweight pollen is carried with it.

The flowers usually emerge early in the spring, before the leaves, so that the leaves do not block the movement of the wind. The pollen is deposited on the exposed feathery stigma of the flower.

Pollination by Water

Some weeds, such as Australian sea grass and pond weeds, are pollinated by water. The pollen floats on water, and when it comes into contact with the flower, it is deposited inside the flower.

Pollination by Deception

Orchids are highly valued flowers, with many rare varieties. They grow in a range of specific habitats, mainly in the tropics of Asia, South America, and Central America. At least 25,000 species of orchids have been identified.

Flowers often attract pollinators with food rewards, in the form of nectar. However, some species of orchid are an exception to this standard: they have evolved different ways to attract the desired pollinators. They use a method known as food deception, in which bright colors and perfumes are offered, but no food. Anacamptis morio, commonly known as the green-winged orchid, bears bright purple flowers and emits a strong scent. The bumblebee, its main pollinator, is attracted to the flower because of the strong scent—which usually indicates food for a bee—and in the process, picks up the pollen to be transported to another flower.

Other orchids use sexual deception. Chiloglottis trapeziformis emits a compound that smells the same as the pheromone emitted by a female wasp to attract male wasps. The male wasp is attracted to the scent, lands on the orchid flower, and in the process, transfers pollen. Some orchids, like the Australian hammer orchid, use scent as well as visual trickery in yet another sexual deception strategy to attract wasps. The flower of this orchid mimics the appearance of a female wasp and emits a pheromone. The male wasp tries to mate with what appears to be a female wasp, and in the process, picks up pollen, which it then transfers to the next counterfeit mate.

Pollination Control Techniques

Pollination Control

Pollination is the process of the transfer of pollen from the anthers of a flower to the stigma of the same flower or another flower. The process of fertilization takes place only after pollination has occurred and allows the flower to develop seeds. Some plant species can be self-pollinated, which occurs when pollen and pistil are from the same plant. Other species are cross-pollinated, which means the pollen and pistil are from different plants. Usually plants need a pollinator for transfer of pollen to the pistil. Some common pollinators are insects, wind, and animals. Plant breeders should have a thorough knowledge of the flower structure as well as the pollination mechanism of the plant species of interest. A flower can be complete, that is when all of the four major parts (petal, sepal, stamen and pistil) are present, or incomplete (when either of the part is missing). A flower can also be perfect if both stamen and pistil are present or imperfect if the flower is unisexual. When a male flower, staminate and a female flower, pistillate are present on the same plant, the plants are called monoecious. In some species, plants have either pistillate or staminate flowers, and they are called dioecious plants. The transfer of genes from one parent plant to another parent plant occurs by cross-pollination, whereas in self-pollination there is only one parent. Cross-pollination promotes heterozygosity. On the other hand, self-pollination promotes homozygosity. Therefore, it becomes very important in any plant breeding experiment to ensure only desired pollen is involved. Hybrid cultivar breeding is one of the most flourishing branches of plant breeding. It is based on the principle of "Heterosis" (hybrid vigor), which is explained as the superiority in vigor and productivity of progeny from the parental genotype. Hybrids are the first generation offsprings produced by controlled crosses between distinct inbred lines of parents. There are three basic steps involved in hybrid cultivar development:

1. Development of inbred parental lines,
2. Crossing of unrelated inbred parent lines to produce the F_1 generation,
3. Producing seeds for distribution.

Hybrid seed production is dependent upon the control of pollination. The first two steps in hybrid cultivar development involve pollination control to get the desired result. Pollination control refers to the technique used to prevent undesired pollination in the plant of interest. Controlled pollination allows the combining of elite lines to produce high quality progeny. Controlling pollination in dioecious plants is relatively easy as male and female flowers occur on separate plants and therefore can be planted away from each other. Pollination control becomes a big problem in monoecious plants. Pollination control is widely divided in to three categories:

1. Mechanical,
2. Chemical,
3. Genetic.

Mechanical Control

Mechanical pollination control refers to any approach by which pollen transfer is mechanically prevented. In monoecious crops, where male and female flowers are at different positions on the plant, whole male flowers can be removed from the plant to control pollination. In corn, this process is called detasseling. It is the most widely used method and is done before pollen shed or before the silks appear. A detasseler is a large machine to remove male flowers; many patents have been obtained for the new design of the machine over the years. Detasseling of corn is carried out in two steps. First, a detasseler is used and then left over tassels are removed manually. There are a few disadvantages associated with the use of a mechanical detasseler, such as requirement of manual labor and foliar damage.

Removal of anthers from the flowers is called emasculation. It is commonly done in hermaphrodite flowers. Manual emasculation is very common in breeding experiments, but it is time consuming, intensive, and expensive. The most common procedure of manual emasculation is with the use of forceps or scissors. In wheat, emasculation is done by scissors and forceps, and is usually done 1-3 days before anthesis. Care must be taken to minimize damage to the stigma, and after the emasculation, the inflorescence is covered with a paper bag. Emasculation of crops like wheat and barley is very labor intensive and tedious, and usually requires skilled workers. In rice, pollen removal is commonly done by clipping and vacuuming the anthers. A lot of precision is needed in rice emasculation by clipping, as a lower cut can cause damage to stigma. Cotton is generally emasculated by forceps and scissors and should be emasculated later in the afternoon. After the emasculation, flowers are covered with a paper bag until pollination. Several techniques have been developed for emasculation of small size flowers. For example, eucalyptus flowers are emasculated by clipping of anthers, but due to the small size of the flower this becomes very labor intensive. Assis et al. developed a method called Artificially Induced Protogyny (AIP). AIP involves cutting off the tip of the mature flower bud just prior to anthesis, so as to remove the stigma and expose the cut surface of the upper style to which the target pollen is applied, without emasculating or isolating the flower. Harbard et al. developed another method for emasculation of eucalyptus flowers called One Stop Pollination (OSP). OSP requires only one visit to the flower to emasculate and immediate pollination of stigmas to induce receptivity, followed by bagging. Horsley et al. compared various methods of eucalyptus controlled pollination and showed that the AIP method is the most promising technique due to high seed yield and lower labor cost whereas OSP method had the lowest genetic contamination, but lower seed set. Horsley et al. reported the development of a new method for the isolation of stigma using sodium alginate gel, and was effective in excluding pollination by external pollen. This method was compared to exclusion bags and proved to be better in controlling pollination, and had lower labor cost.

Bags are also commonly used to exclude pollinators to avoid unwanted pollen. Sometimes, bags are used in addition to emasculation when pollination takes place after a few hours or days. The pore size and the type of fabric used are extremely important in any controlled pollination experiment. Various materials that are used for pollination bags are polyester, polyethylene, butter paper, paraffin paper, parchment paper, plastic, nylon, and kraft paper. Neal and Anderson studied the permeability of four different fabrics: large mesh, small mesh, cotton muslin and filter fabric as pollination bags. They concluded that the fabric with the smallest mesh size (smaller than the pollen grain) was impermeable to the pollen. Cotton muslin also offered a highly significant barrier to wind borne pollen. Pickering showed significant improvement in seed quality and weight in barley by covering the spikes with brown paper bags as compared to glassine and polyethylene. Plant breeders need to make a wise decision when choosing a bag for their experiment. This is affected by whether the plant is pollinated by wind, or insects (small or big), and weather conditions like rain. Some bag materials may trap heat, causing flowers or spikes to reach lethal temperatures. Ball et al. showed that the closely fitted bags made of white paper result in the coolest temperature, whereas a loosely fitted bag made of transparent material gives high temperatures. Synthetic bags have the advantage of drying quickly, whereas cotton muslin and small mesh size bags can retain moisture for a long time. Hot water emasculation is also another commonly used method for deactivating pollen. It is an easy method and does not necessarily require skilled labor. Usually, the inflorescence is allowed to dry after hot water emasculation before pollination. Hot water emasculation is commonly used in sorghum, rice, and sugar cane, and involves soaking of the flower in hot water instead of using forceps and scissors to remove stamens. Mukasa et al. and Otsuka et al. demonstrated that soaking of apical clusters of buckwheat flower in water at 44 °C water for 3 min was successful in controlling pollination and is capable of practical use in tartary buckwheat. Alcohol emasculation is also a commonly used technique for inactivation of pollen. Tysdal and Garl were the first to introduce alcohol emasculation in alfalfa and showed that it was more efficient than the suction method. Sgaier showed that the progeny of alfalfa plants that received suction emasculation as compared to alcohol emasculation were shorter and yielded less dry matter. Alfalfa is one of the most commonly studied plants for alcohol emasculation. Successful emasculation of rice and clover pollen has been achieved by concentrations ranging from 60-76% ethyl alcohol. Another method of emasculation is vacuuming of anthers, which was originally developed by Savage. This method involves an aspirator and high speed suction of pollen from the anthers of sweet clover flowers with a minimum damage to the stigma. Aspirator assisted emasculation was proven to be useful for sugarcane flower, and can potentially be useful for other small flowers. Another method that has been studied for emasculation is chilling or low temperature treatment.

Suneson reported the deactivation of wheat pollen by exposure of stamens to low temperatures of -3.5 °C to 2 °C. Emasculation by chilling is not a commonly used technique due to the adverse effect of low temperature on the whole plant.

Chemical Control

Manyt of the major crops, like wheat and onion, have small bisexual flowers which makes manual emasculation impractical due to intensive labor and time requirements. Use of effective gametocides in inducing male sterility without causing any adverse effects on plant growth and environment is highly desirable to overcome problems related to mechanical emasculation, as well as genetic pollination control. The effects of gametocides may range from irregular meiosis to dysfunctional pollen grain formation. The advantages of using chemical control for inducing male sterility are that any parent can be used as a female, and there is no requirement for a fertility restorer gene.

A variety of chemicals with gametocidal properties are used to produce male sterility temporarily in crops. Dalapon, estrone, ethephon, and generis are a few of the examples of the chemicals that are used commercially to induce male sterility. Ethephon (2 chloroethyl phosphonic acid) is known to advance the formation of pistillate flowers and delay staminate flower formation. Lower and Miller studied male sterility in cucumber and concluded Ethrel (ethephon) to be very effective in causing delayed growth of staminate flower and therefore it has great potential in being useful in plant hybridization experiments. Rowell and Miller showed that ethephon spray induced male sterility in wheat, and that the percentage of male sterility depended on the concentration of ethephon and the stage of the plant. Later, Rowell and Miller concluded that the maturity stage of the wheat plant is critical for ethephon spray (early boot to mid boot stage), as ethephon can also affect female fertility. Jan et al. conducted an experiment on the use of three gametocides (RH531, RH532 and RH2956) which involved the timing and rate of application on wheat. They concluded that RH532 and RH2956 were potential candidates for maximum male sterility without affecting female fertility. They also concluded that the application of gametocides at booting stage was most effective. Chauhan et al. reported the potential use of benzotriazole as a hybridizing agent by causing pollen sterility in pepper, cotton, and radish. Chauhan and Vandana compared different hybridizing agents for Brassica juncea, and reported that benzotriazole exhibited complete pollen sterility as compared to 92-96% pollen sterility by ethephon. Complete male sterility in maize, cotton and sugar beet can be achieved by the use of gibberellins and sodium 2, 3-dichlorobutyrate (FW 450). In studies on the gametocidal effect of 2, 3-dichloroisobutyrate (FW 450) on cotton and soybean respectively, Pate and Duncan and Starnes and Hadley concluded that this chemical was not able to selectively induce male sterility and hence reduced female fertility as well as resulting in lower seed set. Ali et al. studied the efficacy of twenty different oxinalates in comparison to methyl arsenate in inducing male sterility in Oryza sativa, and concluded that the halogen substituted oxinalates were the most effective. Specifically, ethyl 4′fluoroxanilate and ethyl 4′bromooxanilate induced the highest pollen sterility with the least phytotoxicity. Ethyloxanilates, the new class of gametocides, have shown promising results in inducing male sterility without affecting female fertility and yield. DPX 3778, a triethanolamine salt, is a chemical hybridizing agent known to control pollination by preventing anther dehiscence with no effect on male and female fertility in corn, wheat, oats, and pearl millet. The most effective time of application has been found to be early boot stage. Theurer conducted a study on the effect of DPX 3778 on sugar beet and reported that DPX 3778 resulted in reduced anther dehiscence, but also resulted in failure of flower opening, making DPX ineffective in breeding sugar beet.

A commonly observed problem with the use of chemicals for male sterility is the effect on female fertility. Therefore, the use of chemical spray becomes undesirable in many breeding experiments. Moreover, gametocides are expensive and can potentially be a pollutant in the environment.

Genetic Control

In some plant species, certain genes occur naturally that contribute to genetic control over pollination. This occurs mainly by two mechanisms:

1. Male sterility, in which the male reproductive system cannot produce viable pollen.
2. Self-incompatibility (SI), in which viable pollen from a flower is not able to fertilize the same flower or a flower with the same genotype as the pollen.

In male sterile plants, anthers or pollen are non functional, which eliminates the need for any sort of emasculation. Male sterility can be divided into 3 categories: cytoplasmic, nuclear, and cytoplasmic-genetic.

Cytoplasmic male sterility (CMS) is controlled by the mitochondrial gene and inherited as a dominant, maternally transmitted trait. CMS has been found in corn, sorghum, and flax. CMS is used in many breeding programs as the progeny are male sterile. CMS lines can be used for hybrid production when CMS mutants are available for the crop, when restorer genes to restore fertility are present, when seed is of interest, and when CMS is not associated with yield penalty. It has been reported that in agronomic crops, CMS hybrids are associated with disease susceptibility, and sterility is unstable. Nuclear male sterility is also called genic male sterility (GMS). In GMS the plants fail to develop pollen or form abnormal anthers. GMS is commonly known to occur in tomato, potato, soybean, cotton, and barley. GMS usually occurs due to a single recessive gene "ms", dominant allele of which "Ms" produces normal anthers and pollen. Therefore, whenever a genic male sterile (msms) crosses with a fertile heterozygous (Msms), 50% of the progeny will be male sterile and thus pure male sterile population can never be achieved. Due to this type of constraint GMS becomes uneconomical, as male sterility cannot be identified before flowering, and the chances of pollen contamination increases. Molecular marker assisted selection can be used to identify the male steriles, but development of markers associated with the locus is needed. One way to avoid this type of situation is by clonal propagation, but this type of propagation is also limited to some ornamental and vegetables. Another approach can be the use of GMS as pollination control when GMS is conditioned by a dominant gene. This has been done in cotton and wheat, but in this case, the F1 hybrid turned out to be sterile and therefore not desired. Mariani et al. reported the first transgene to confer GMS that could facilitate hybrid seed production. Subsequently, several GMS transgenes have been reported. Many transgenes result in male sterility by disrupting tissue specific gene expression of the protein that is needed for production of functional pollen. The introduction of one or more genes that can alter levels of metabolites, like amino acids and sugars that are needed for pollen formation can also be used to cause male sterility. Male sterility can also be achieved by natural or induced mutation, and then the wild type might have a possibility of having the restorer gene.

Cytoplasmic-genetic male sterility is caused by cytoplasmic genes. It is restored by a fertility restoring (Rf) gene (nuclear gene), which results in the development of normal anthers and pollen. This system has been widely used in hybrid seed production in onion, sorghum, safflower, corn, and sugarbeet, as it overcomes the problems of GMS and CMS. This system imparts a convenience to breeders to control the sterility expression by manipulating the gene–cytoplasm combinations in any selected genotype, and hence becomes highly desirable.

Self incompatibility prevents inbreeding depression. SI is commonly found in some cultivated crops like rye, potato, alfalfa, clover, pearl millet, and sugarbeets. The incompatibility reaction is controlled by the S locus. This locus has multiple alleles, each of which contributes in determining the reaction between pistil and pollen. SI can be either heteromorphic, in which differences in the length of style and stamen causes incompatibility, or homomorphic. Homomorphic is further divided into gametophytic, when the genotype of the pollen determines the fate, or sporophytic when the genotype of the sporophyte that produces the pollen determines the incompatibility. SI promotes heterozygosity and therefore is desirable. When it is desired to overcome SI to promote homozygosity, techniques like removal of stigma surface and bud pollination can be used. This mechanism of SI is widely used in Brassica and Raphanus for production of single-cross hybrid seeds.

Pollination control is an important aspect of plant breeding and hybrid seed production. Many systems for pollination control have been studied over the decades in search of an ideal technique to control pollination. The choice of the type of pollination control used is largely dictated by the crop of interest. Mechanical pollination control is still most widely used method but is very labor and time intensive. Therefore becomes uneconomical as well as limits the number of crosses that can be made. Chemical control is promising, but not widely used due to the adverse effect on female fertility, yield, and potentially, the environment. Genetic control is the most desirable method, as it does not involve any additional steps towards pollination, but it is not present in all crops. Genetic control has also been observed to be linked to disease susceptibility and lower yields. Much advancement have been made over the years to come up with a better pollination control method which can facilitate plant breeding experiments, but still more research is needed to come up with practical methods.

Fertilization

Fertilization is the fusion of male and female gametes. It results in the formation of single-celled zygote.

Germination of Pollen Grain

The pollen grain develops pollen tube through the style and reaches the ovary. It enters the ovule from a small opening. Pollen tube carries the male gamete. Male gamete fuses with the female gamete to form the zygote.

Post-fertilization Change

Following changes take place in the flower after fertilization. The fertilized zygote grows into an embryo. The embryo has two parts – plumule which grows into shoot and radical which grows into the root.

- The ovary grows into fruit.

- The other parts of the flower fall off.
- The ovule develops into a seed.

Double Fertilization

Double fertilization is a complex fertilization mechanism of angiosperms which involves the fusion of a female gametophyte with two male gametes. One male gamete fuses with the female egg whereas the second sperm fuses with two polar nuclei to give rise to a triploid body that forms an endosperm.

Double Fertilization in Angiosperms

Angiosperms are flower-bearing plants surrounding us and these flowers are their reproductive units which include male and female reproductive organs. Each contains gametes – sperm and egg cells respectively. Pollination helps the pollen grains to reach stigma via style. Two sperm cells enter the ovule-synergid cell. This proceeds to fertilization.

In angiosperms, fertilization results in two structures namely zygote and endosperm; thus fertilization is known as double fertilization. By definition, double fertilization is a complex fertilization method where out of two sperm cells, one fuses with the egg cell and other fuses with two polar nuclei which result in diploid (2n) zygote and triploid (3n) primary endosperm nucleus (PEN) respectively.

Since endosperm is a product of the fusion of three haploid nuclei, it is called triple fusion. Eventually, the primary endosperm nucleus develops into the primary endosperm cell (PEC) and then into the endosperm. The zygote becomes an embryo after numerous cell divisions.

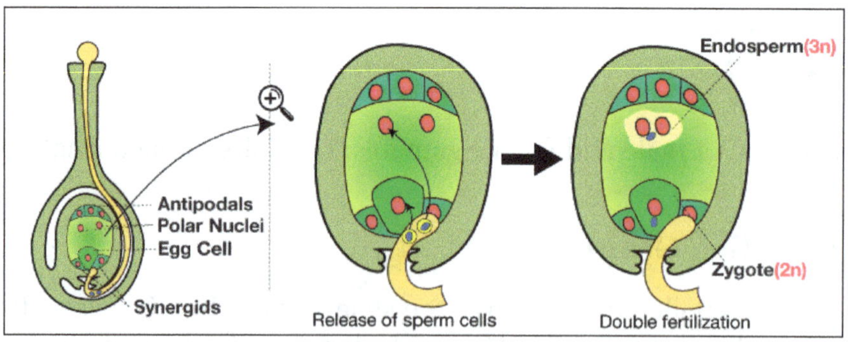

Development of Embryo

Once fertilization is done embryonic development starts and no more sperm can enter the ovary. The fertilized ovule develops into a seed and ovary tissues develop as fleshy fruit which encloses the seed.

The first stage of embryonic development includes division of the zygote into the upper terminal cell and lower basal cell.

The basal cell develops into suspensor which helps in nutrition transport to the growing embryo. The terminal cell develops into proembryo. Further embryonic development into seed takes in an interval of time.

Significance of Double Fertilization

The significance of double fertilization is mentioned below:

Produces a Food Supply for the seed

An embryo is produced within the seed during the process of double fertilization. This procedure is also used as an additional food source by some plants.

Provides Extra Protection to the Species

Double fertilization increases the life span of a flowering plant. There are two male gametes. Therefore the chances of them fusing with the female gametes are very high. The endosperm formed after fertilization stores nutrients. So, the plant does not have to allocate energy-intensive resources for the unfertilized egg.

Energy Conservation

When the embryo starts coming into shape, it starts conserving energy due to double fertilization. The essential supply is found in the seed itself.

Allows Rapid Seed Development

Since two male gametes are involved in double fertilization, the development of seed is more rapid as compared to other forms of reproduction. The nucleus of the endosperm divides quickly to structure the nutrient-rich tissue, which inturn enhances the development of embryo.

Plant Embryogenesis

In plants, the term embryogenesis covers development from the time of fertilization until dormancy occurs. The basic body plan of the sporophyte is established during embryogenesis; however, this plan is reiterated and elaborated after dormancy is broken. The major challenges of embryogenesis are:

1. To establish the basic body plan. Radial patterning produces three tissue systems, and axial patterning establishes the apical-basal (shoot-root) axis.
2. To set aside meristematic tissue for postembryonic elaboration of the body structure (leaves, roots, flowers, etc.).
3. To establish an accessible food reserve for the germinating embryo until it becomes autotrophic.

Embryogenesis is similar in all angiosperms in terms of the establishment of the basic body plan. There are differences in pattern elaboration, however, including differences in the precision of cell division patterns, the extent of endosperm development, cotyledon development, and the extent of shoot meristem development.

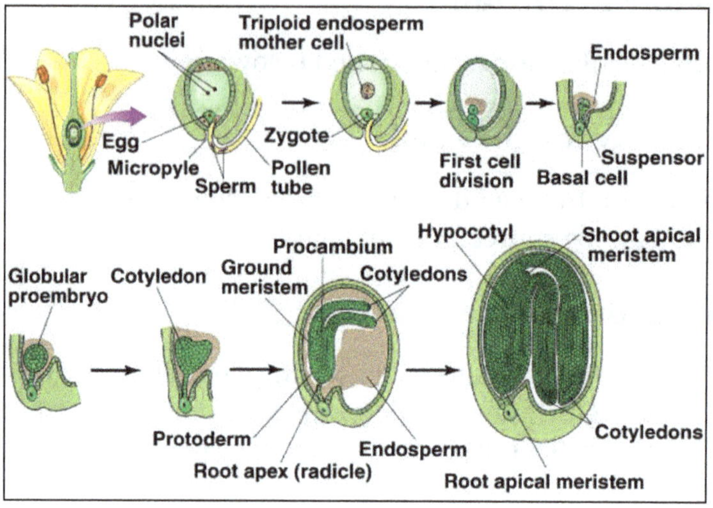

Angiosperm embryogenesis.

Polarity is established in the first cell division following fertilization. The establishment of polarity has been investigated using brown algae as a model system. The zygotes of these plants are independent of other tissues and amenable to manipulation. The initial cell division results in one smaller cell, which will form the rhizoid (root homologue) and anchor the rest of the plant, and one larger cell, which gives rise to the thallus (main body of the sporophyte). The point of sperm entry fixes the position of the rhizoid end of the apical-basal axis. This axis is perpendicular to the plane of the first cell division. F-actin accumulates at the rhizoid pole. However, light or gravity can override this fixing of the axis and establish a new position for cell division.Once the apical-basal axis is established, secretory vesicles are targeted to the rhizoid pole of the zygote. These vesicles contain material for rhizoid outgrowth, with a cell wall of distinct macromolecular composition. Targeted secretion may also help orient the first plane of cell division. Maintenance of rhizoid versus thallus fate early in development depends on information in the cell walls. Cell wall information also appears to be important in angiosperms.

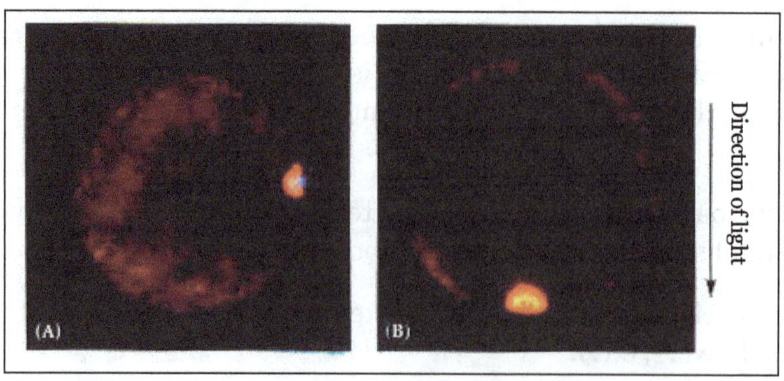

Axis formation in the brown alga Pelvetia compressa. (A) An F-actin patch (orange) is first formed at the point of sperm entry (the blue spot marks the sperm pronucleus). (B) Later, light was shone in the direction of the arrow. The sperm-induced axis was overridden, and an F-actin patch formed on the dark side, where the rhizoid will later form.

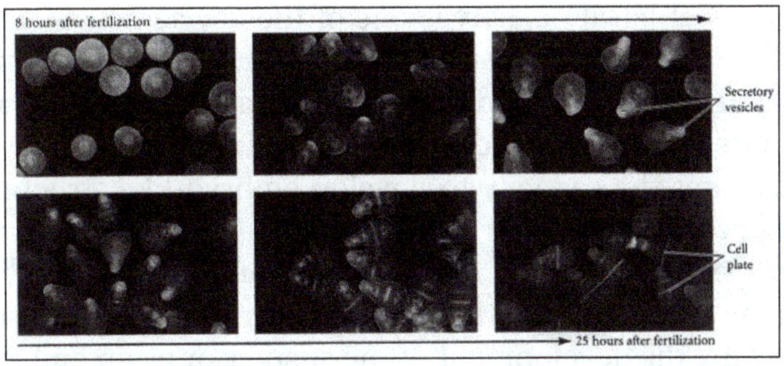

Asymmetrical cell division in brown algae. Time course from 8 to 25 hours after fertilization, showing algal cells stained with a vital membrane dye to visualize secretory vesicles, which appear first, and the cell plate, which begins to appear about halfway through this sequence.

The basic body plan of the angiosperm laid down during embryogenesis also begins with an asymmetrical cell division, giving rise to a terminal cell and a basal cell. The terminal cell gives rise to the embryo proper. The basal cell forms closest to the micropyle and gives rise to the suspensor. The hypophysis is found at the interface between the suspensor and the embryo proper. In many species it gives rise to some of the root cells. (The suspensor cells divide to form a filamentous or spherical organ that degenerates later in embryogenesis.) In both gymnosperms and angiosperms, the suspensor orients the absorptive surface of the embryo toward its food source; in angiosperms, it also appears to serve as a nutrient conduit for the developing embryo. Culturing isolated embryos of scarlet runner beans with and without the suspensor has demonstrated the need for a suspensor through the heart stage in dicots. Embryos cultured with a suspensor are twice as likely to survive as embryos cultured without an attached suspensor at this stage. The suspensor may be a source of hormones. In scarlet runner beans, younger embryos without a suspensor can survive in culture if they are supplemented with the growth hormone gibberellic acid.

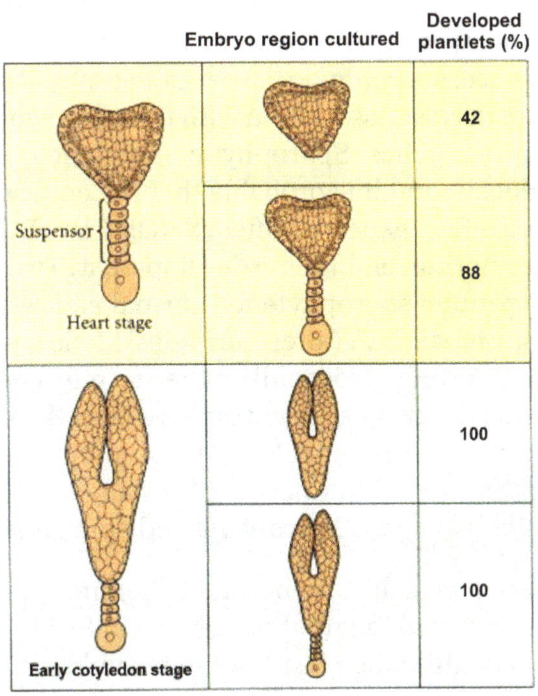

Role of the suspensor in dicot embryogenesis. Culturing scarlet runner bean embryos with and without their suspensors has demonstrated that the suspensor is essential at the heart-shaped stage, but not later.

As the establishment of apical-basal polarity is one of the key achievements of embryogenesis, it is useful to consider why the suspensor and embryo proper develop unique morphologies. Here the study of embryo mutants in maize and Arabidopsis has been particularly helpful. Investigations of suspensor mutants (sus1, sus2, and raspberry1) of Arabidopsis have provided genetic evidence that the suspensor has the capacity to develop embryo-like structures. In these mutants, abnormalities in the embryo proper appear prior to suspensor abnormalities. Earlier experiments in which the embryo proper was removed also demonstrated that suspensors could develop like embryos. A signal from the embryo proper to the suspensor may be important in maintaining suspensor identity and blocking the development of the suspensor as an embryo. Molecular analyses of these and other genes are providing insight into the mechanisms of communication between the suspensor and the embryo proper.

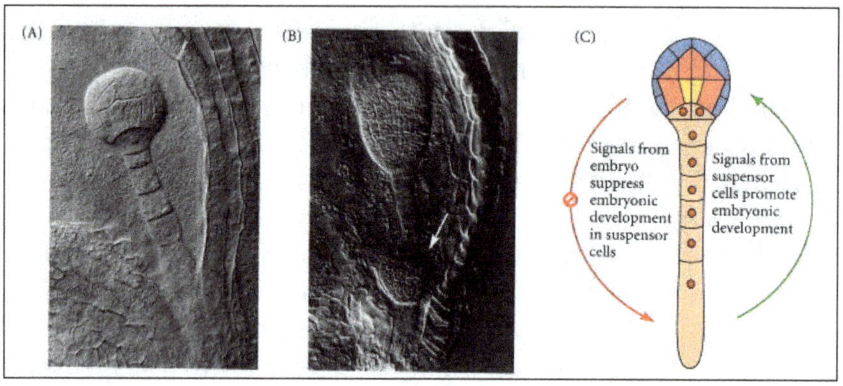

The SUS gene suppresses embryonic development in the suspensor. (A) Wild-type embryo and suspensor. (B) sus mutant with suspensor developing like an embryo (arrow). (C) Model showing how the embryo proper suppresses embryonic development in the suspensor and the suspensor provides feedback information to the embryo.

Maternal effect genes play a key role in establishing embryonic pattern in animals. The role of extrazygotic genes in plant embryogenesis is less clear, and the question is complicated by at least three potential sources of influence: sporophytic tissue, gametophytic tissue, and the polyploid endosperm. All of these tissues are in close association with the egg/zygote. Endosperm development could also be affected by maternal genes. Sporophytic and gametophytic maternal effect genes have been identified in Arabidopsis, and it is probable that the endosperm genome influences the zygote as well. The first maternal effect gene identified, SHORT INTEGUMENTS 1 (SIN1), must be expressed in the sporophyte for normal embryonic development. Two transcription factors (FBP7 and FBP11) are needed in the petunia sporophyte for normal endosperm development. A female gametophytic maternal effect gene, MEDEA (after Euripides' Medea, who killed her own children), has protein domains similar to those of a Drosophila maternal effect gene. Curiously, MEDEA is in the Polycomb gene group, whose products alter chromatin, directly or indirectly, and affect transcription. MEDEA affects an imprinted gene that is expressed by the female gametophyte and by maternally inherited alleles in the zygote, but not by paternally inherited alleles. How significant maternal effect genes are in establishing the sporophyte body plan is still an unanswered question.

Radial and axial patterns develop as cell division and differentiation continue. The cells of the embryo proper divide in transverse and longitudinal planes to form a globular stage embryo with several tiers of cells. Superficially, this stage bears some resemblance to cleavage in animals, but the nuclear/cytoplasmic ratio does not necessarily increase. The emerging shape of the embryo depends on regulation of the planes of cell division and expansion, since the cells are not able to move and reshape the embryo. Cell division planes in the outer layer of cells become restricted, and this layer, called the protoderm, becomes distinct. Radial patterning emerges at the globular stage as the three tissue systems (dermal, ground, and vascular) of the plant are initiated. The dermal tissue (epidermis) will form from the protoderm and contribute to the outer protective layers of the plant. Ground tissue (cortex and pith) forms from the ground meristem, which lies beneath the protoderm. The procambium, which forms at the core of the embryo will give rise to the vascular tissue (xylem and phloem), which will function in support and transport. The differentiation of each tissue system is at least partially independent. For example, in the keule mutant of Arabidopsis, the dermal system is defective while the inner tissue systems develop normally.

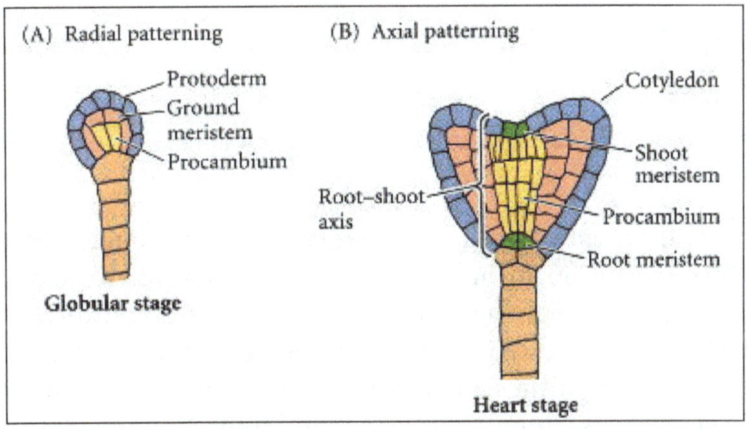

Radial and axial patterning. (A) Radial patterning in angiosperms begins in the globular stage and results in the establishment of three tissue systems. (B) The axial pattern (shoot-root axis) is established by the heart stage.

The globular shape of the embryo is lost as cotyledons ("first leaves") begin to form. Dicots have two cotyledons, which give the embryo a heart-shaped appearance as they form. The axial body plan is evident by this heart stage of development. Hormones (specifically, auxins) may mediate the transition from radial to bilateral symmetry. In monocots, such as maize, only a single cotyledon emerges.

In many plants, the cotyledons aid in nourishing the plant by becoming photosynthetic after germination (although those of some species never emerge from the ground). In some cases—peas, for example—the food reserve in the endosperm is used up before germination, and the cotyledons serve as the nutrient source for the germinating seedling. Even in the presence of a persistent endosperm (as in maize), the cotyledons store food reserves such as starch, lipids, and proteins. In many monocots, the cotyledon grows into a large organ pressed against the endosperm and aids in nutrient transfer to the seedling. Upright cotyledons can give the embryo a torpedo shape. In some plants, the cotyledons grow sufficiently long that they must bend to fit within the confines of the seed coat. The embryo then looks like a walking stick. By this point, the suspensor is degenerating.

The shoot apical meristem and root apical meristem are clusters of stem cells that will persist in the postembryonic plant and give rise to most of the sporophyte body. The root meristem is partially derived from the hypophysis in some species. All other parts of the sporophyte body are derived from the embryo proper. Genetic evidence indicates that the formation of the shoot and root meristems is regulated independently. This independence is demonstrated by the dek23 maize mutant and the shootmeristemless (STM) mutant of Arabidopsis, both of which form a root meristem but fail to initiate a shoot meristem. The STM gene, which has been cloned, is expressed in the late globular stage, before cotyledons form. Genes have also been identified that specifically affect the development of the root axis during embryogenesis. Mutations of the HOBBIT gene in Arabidopsis, for example, affect the hypophysis derivatives and eliminate root meristem function.

The shoot apical meristem will initiate leaves after germination and, ultimately, the transition to reproductive development. In Arabidopsis, the cotyledons are produced from general embryonic tissue, not from the shoot meristem. In many angiosperms, a few leaves are initiated during embryogenesis.

Sporogenesis

Plant reproduction is initiated by the specification of sporocytes that form haploid spores through meiosis.

The life cycle of plants alternates between a diploid sporophytic generation and a haploid gametophytic generation. The transition from sporophyte to gametophyte requires the formation of haploid cells (spores) by meiosis, a process called sporogenesis. In flowering plants, sporogenesis is initiated in both male and female reproductive tissues by the differentiation of sub-epidermal cells in the anther and ovule primordia, called archesporial cells (ACs), into pre-meiotic cells called sporocytes. Both male sporocytes (microsporocytes) and female sporocytes (megasporocytes) will then undergo meiosis to give rise to the microspores and megaspores, in the male and female organs, respectively. The specification of the sporocytes represents a critical step in the process of sporogenesis. The SPOROCYTELESS/NOZZLE (SPL/NZZ) gene which encodes a nuclear protein was identified as a key regulator of this stage of sporogenesis in Arabdiopsis. In spl/nzz mutants, development of both male and female reproductive organs proceeds normally until the ACs are formed, but the ACs fail to differentiate into sporocytes. Subsequently, SPL was found to be the target of regulation by the floral homeotic gene AGAMOUS (AG). AG is a floral organ identity gene required for the specification of stamens and carpels in the third and fourth whorls of the flower. AG encodes a transcription factor of the MADS-box family expressed in stamen and carpel primordia, and is required for the activation of SPL expression. In anthers, there are additionally two leucine-rich repeat receptor-like protein kinases, BARELY ANY MERISTEM1 (BAM1) and BAM2, that act in a regulatory loop with SPL to promote somatic cell types and to restrict SPL expression to the inner region of the locule.

Despite the identification of SPL/NZZ 15 years ago, and its central role in promoting the reproductive transition to the gametophytic generation, the molecular basis of gene regulation by SPL has remained unknown. A new study by Wei et al. has now revealed the molecular mechanism by which SPL acts to control gene expression. Wei et al. show that the C-terminal end of SPL protein contains an EAR motif, which presents in many transcriptional repressors, and that an intact EAR motif is required for rescue of a spl mutation. Furthermore, SPL interacts with co-repressors of the TPL/TPR family through its EAR motif. Expression of a mutant protein SPLmEAR (SPL with mutated EAR motif) results in a spl mutant phenotype due to competition with endogenous SPL. This dominant negative mutant genetically interacts with the tpl-1 mutant which is defective for the TPL1 gene. Moreover, in tpl-1 homozygous mutant ovules, ACs failed to differentiate into megasporocytes. These results suggest that SPL regulates megasprogenesis through the recruitment of the TPL1 co-repressor, forming a repressor complex.

Which genes might be the target of regulation by this repressor complex? By using yeast 2-hybrid screens, Wei et al. show that the N-terminus of SPL interacts with transcription factors of the TCP (TEOSINTE BRANCHED1/CYCLOIDEA/PROLIFERATING CELL FACTOR) family. Moreover, overexpression of SPL resulted in phenotypes resembling loss-of-function mutants in the class II TCP genes that are required for leaf development, suggesting that ectopic SPL interferes with TCP functions in the leaf. Wei et al. also found that class II TCPs are expressed in the ovule, consistent with their roles in ovule development. Overexpression of these TCPs results in ovule phenotypes similar to spl mutants, including suppression of AC differentiation into megasporocytes. Furthermore, protein-protein interaction experiments showed that SPL likely serves as a bridge between TCPs and TPL/TPR proteins. In shoot meristems, the CUC (Cup Shaped Cotyledon) genes that promote meristem formation are the presumptive targets of class II TCPs. Class II TCPs in the shoot act to restrict undifferentiated or meristematic cell fate through downregulation of CUC genes at organ boundaries, and thus promote organ differentiation. The target genes of TCPs in ovules and anthers are yet to be identified, but may have a similar role to those in shoots in preventing differentiation. The findings by Wei et al. provide for the first time insights into the molecular mechanism for the initiation of sporogenesis in a flowering plant. The floral homeotic gene AG activates SPL expression in the archesporial cells of anthers and ovules. SPL protein associates with class II TCPs and recruits TPL/TPR co-repressors to suppress target genes of the TCP transcription factors. It can be concluded that SPL likely promotes sporogenesis by acting in concert with TCPs and TPL/TPRs to suppress negative regulators of sporocyte differentiation.

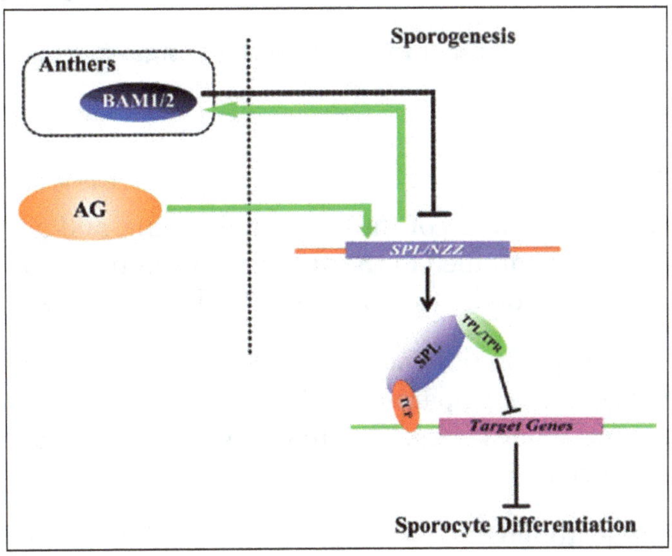

Sporogenesis in Arabidopsis. Schematic representation, based on the results of Wei et al., and incorporating the previously known genetic interactions between genes involved in sporogenesis in Arabidopsis. Arrows indicate positive regulation, and black T-bars indicate negative regulation. Dotted lines show indirect interactions.

This proposed mechanism for regulating sporogenesis in Arabidopsis shows some remarkable parallels to that in fungi. Similar to plants, fungi commonly have alternating haploid and diploid generations, and the formation of haploid spores constitutes an important step in their reproductive strategies.

Gametogenesis

Gametogenesis is a biological process by which haploid male and female gametes are formed. This occurs in both plants and animals. In higher plants, there are two stages that are involved- sporogenesis and gametogenesis. Sporogenesis is the formation of spores whereas gametogenesis is the formation of gametes.

Gametogenesis: Formation of Male Gametes

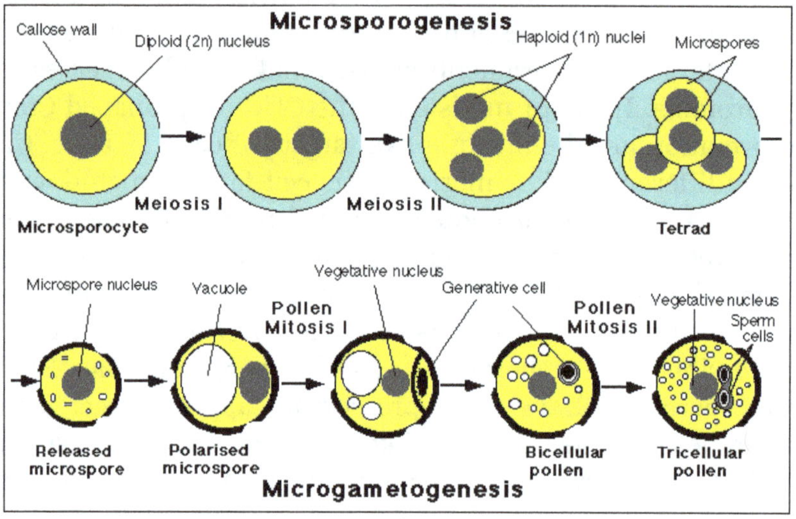

Gametogenesis leading to the formation of male gametes in angiosperms occurs in two stages:

- Microsporogenesis
- Microgametogenesis

The androecium of the plant is made up of the stamen, filament, and anthers. The pollen grains that carry the male gametes are contained in the anthers. The anther is anatomically divided into lobes and the lobes are further divided into chambers called microsporangia.

Pollens present in these microsporangia. The microsporangia are also known as pollen sac. It is said that the pollen sac develops from a parent cell known as the archesporial cell. This archesporial cell divides into outer and inner layers. The inner layer forms the sporogenous tissue or cell that forms the pollen mother cell.

This sporogenous tissue is surrounded by the tapetum layer which provided nourishment to the pollen or microspore mother cell present in the sporogenous tissue. The microspore mother cell is diploid and divides meiotically to form haploid microspores or pollen. This describes the process of microsporogenesis.

The pollen grain is the haploid cell that is made up of two layers: the outer exine layer which is derived from the tapetum and the inner layer called intine. The exine is present all over the pollen grain except in one small part from where the pollen tube emerges post pollination. This small part is known as the germ pore.

The pollen grain divides into two halves- the small generative nucleus and the larger vegetative nucleus. The generative cell gives rise to two male nuclei whereas the vegetative cell gives rise to the pollen tube. This explains the process of microgametogenesis.

Gametogenesis: Formation of Female Gametes

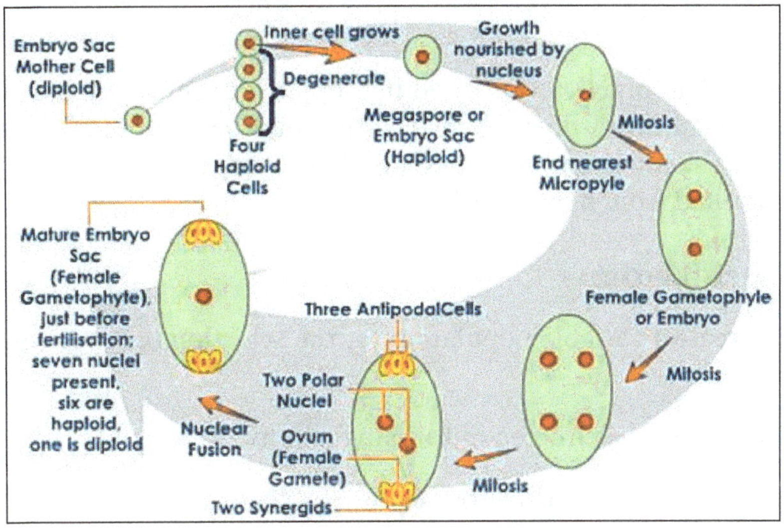

Gametogenesis in angiosperms to form the female gametes, like the male gametes, occurs in two stages:

- Megasporogenesis
- Megagametogenesis

The word 'mega-' is used instead of 'micro-' in females because the female gametes are said to be larger in size as compared to the male gametes.

The ovules are present inside the ovary in multiple lobes. A cell in the ovule differentiates into a megaspore mother cell. The megaspore mother cell is diploid. This megaspore mother cell undergoes meiosis to form 4 haploid megaspores. 3 of the 4 megaspores degenerate and only one megaspore is left in each ovule. This process is known as megasporogenesis.

This megaspore nucleus now begins to divide mitotically to form 8 nuclei. 6 of the 8 nuclei migrate to opposing poles (3 each) while two nuclei remain at the center. The nuclei that remain at the center are known as polar nuclei. These polar nuclei fuse to form the secondary nucleus. The megaspore matures into an embryo sac. This entire process is called as megagametogenesis.

Hybridization

Individual produced as a result of cross between two genetically different parents is known as hybrid. The natural or artificial process that results in the formation of hybrid is known as hybridization. In other words, the production of a hybrid by crossing two individuals of unlike genetical constitution is known as hybridization. Hybridization is an important method of combining characters of different plants. Hybridization does not change genetic contents of organisms but it produces new combination of genes.

The first natural hybridization was recorded by Cotton Mather (1716) in corn. The first artificial interspecific plant hybrid was produced by Thomas Fairchild in 1717. It is commonly known as 'Fairchild Mule'.

Hybridization was first of all practically utilized in crop improvement by German botanist Joseph Koerauter in 1760. Mendel onward, the hybridization had become the key method of crop improvement. Today, it is the most common method of crop improvement, and the vast majority of crop varieties have resulted from hybridization.

Objectives of Hybridization

- To artificially create a variable population for the selection of types with desired combination of characters.
- To combine the desired characters into a single individual.
- To exploit and utilize the hybrid varieties.

Types of Hybridization

Hybridization may be of following types:

- Intra-varietal hybridization: The crosses are made between the plants of the same variety.
- Inter-varietal or Intraspecific hybridization: The crosses are made between the plants belonging to two different varieties.
- Interspecific hybridization or intragenric hybridization: The crosses are made between two different species of the same genus.
- Introgressive hybridization: Transfer of some genes from one species into the genome of the other species is known as introgressive hybridization. The crosses between different species of the same genus or different genera of the same family are also known as distant hybridization or wide crossing. Such crosses are called distant crosses.

Procedure of Hybridization

Selection of Parents

The selection of parents depends upon the aims and objectives of breeding. Parental plants must be selected from the local areas and are supposed to be the best suited to the existing conditions.

Selfing of Parents or Artificial Self-pollination

It is essential for inducing homozygosity for eliminating the undesirable characters and obtaining inbreeds.

Emasculation

It is the third step in hybridization. Inbreeds are grown under normal conditions and are emasculated. Emasculation is the removal of stamens from female parent before they burst and shed their pollens.

It can be defined as the removal of stamens or anthers or the killing of the pollen grains of a flower without affecting in any way the female reproductive organs. Emasculation is not required in unisexual plants but it is essential in bisexual or self-pollinated plants.

Various methods used for emasculation are:

- Hand emasculation or forceps or scissor method: This method is generally used in those plants which have large flowers. In this method the corolla of the selected flowers is opened and the anthers carefully removed with the help of fine-tip forceps.

Following are the important precautions while performing this method:

- Flowers should be selected at proper stage.
- Stigma should be receptive and anthers should not have dehisced.
- All the anthers should be removed from the flowers without breaking.
- Stigma and ovary of the flower should not be damaged.

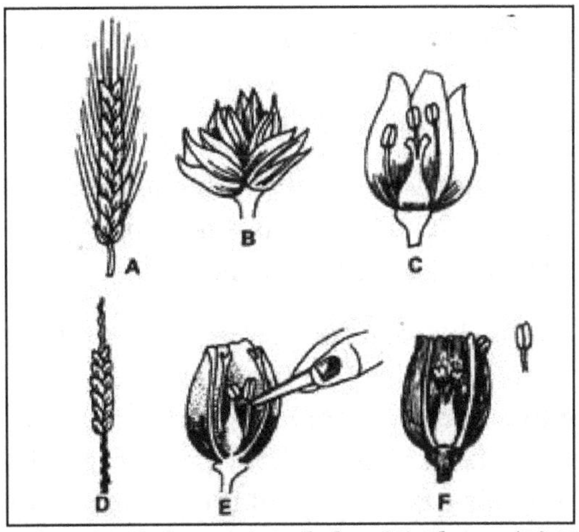

(A-F) Emasculation in wheat. (A) Spike of spikelets, (B) Spikelet, (C) Floret, (D) Upper and lower spikelets removed, awus removed, upper portion of florets cut, (E) Anthers removed with the help of fine-tip forceps. (F) Removed anther.

- Hot water treatment: Removal of stamens with the help of forceps is very difficult in minute flowers. In such small hermaphrodite flowers (e.g., Bajra, Jowar) emasculation is done by dipping the flowers in hot water for a certain duration (1-10 minutes) of time.

The time varies from species to species. This method Is based on the fact that gynoecia can withstand the hot temperature at which the anthers are killed. In this method an equipment is used which is placed on a simple heavy stand.

It consists of a cylindrical metallic container of 60 cm length, with one hole of 5 cm to 16 cm diameter on one end to pass over a bajra or jowar head. After inserting the panicle inside the container a cork is fitted in the hole to close it.

A 35 cm long rubber tube or belt is stretched over the side of the container, and when in use this tube is tied around the peduncle of the head. To measure the temperature, in the upper side of the container a thermometer is placed. In the field water is carried in a thermos jug.

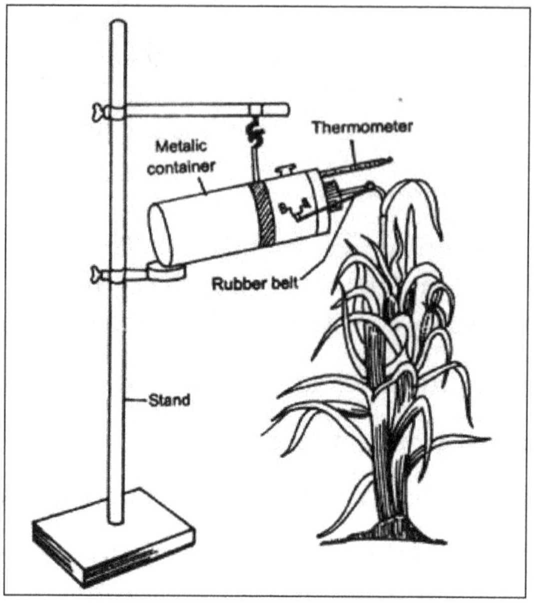

Hot water equipment for emasculation

The panicle is inserted in the container prior to blooming for a particular duration of time. It has been observed that pollen grains of rice are killed by immersing the inflorescence for 5 to 10 minutes in the hot water maintained at 40-44°C in a thermos flask.

- Cold water treatment: Like hot water cold water also kills pollen grains without damaging the gynoecium. In rice 0-6°C temperature is maintained to kill the pollen grains. This method is less effective than hot water treatment.
- Alcohol treatment method: This method is not commonly used for emasculation because duration of treatment is an important factor since a very short duration is required failing which even the gynoecium may be damaged. Flowers or inflorescences are immersed in alcohol of a suitable concentration for a brief period. In alfa-alfa, a treatment of even 10 seconds with 57 % alcohol is sufficient to kill the pollen grains.
- Suction method: It is a mechanical method and is suitable for the crops having minute flowers. In this method the amount of pressure is applied in such a way that only anthers are sucked out and other parts of the flower like gynoecium remain intact. However, in this method 10-15% self pollination takes place. It is one of the major drawback of this method.
- Male sterility or self-incompatibility method: Emasculation option can be eliminated by the use of male-sterile plants, In some self-pollinated plants for example, Sorghum, Onion, Barley etc. anthers are sterile and do not produce any viable pollens! Similarly self-incompatibility may also be used to avoid emasculation.
- Chemical gametocides: Certain chemicals are capable of causing male sterility, when sprayed before flowering e.g., 2, 4-D, naphthalene acetic acid (NAA), maleic-hydrazide (MA), tribenzoic acid etc. FW450 in cotton may be used for bringing about emasculation.

Bagging

It is the fourth step and completed with emasculation. The emasculated flower or inflorescence is immediately bagged to avoid pollination by any foreign pollen. The bags may be made of paper, butter paper, glassine or fine cloth. Butter paper or vegetable parchment bags are most commonly used.

The bags are tied to the base of the inflorescence or to the stalk of the flower with the help of thread, wire or pins. The bagging is done with the emasculation in bisexual plants and before the stigma receptivity and dehiscence of the anthers in unisexual plants. Both male and female flowers are bagged separately to prevent contamination in male flowers and cross-pollination in female flowers.

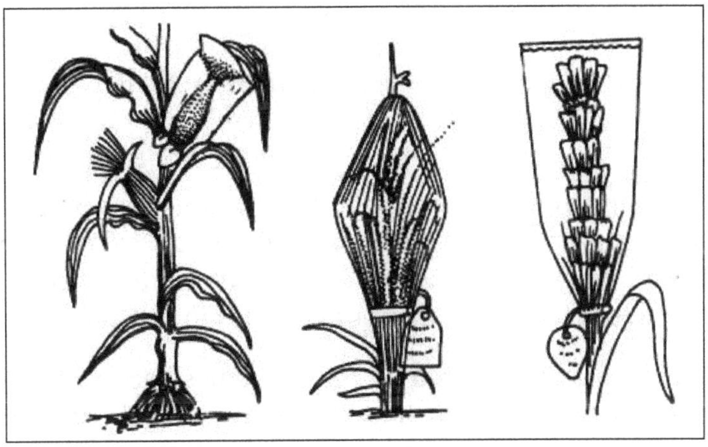

Different method of bagging

Tagging

The emasculated flowers are tagged just after bagging. Generally circular tags of about 3 cm or rectangular tags of about 3 x 2 cm are used. The tags are attached to the base of flower or inflorescence with the help of thread.

The information on tag must be as brief as possible but complete bearing the following information:

- Number referring to the field record
- Date of emasculation
- Date of crossing
- Name of the female parent is written first followed by a cross sign (x) and then the male parent, e.g., C x D denotes that C is the female parent and D is the male parent

Crossing

It is the sixth step. It can be defined as the artificial cross-pollination between the genetically unlike plants. In this method mature, fertile and viable pollens from the male parent are placed on the receptive stigma of emasculated flowers to bring about fertilization.

Pollen grains are collected in petridishs (e.g., Wheat, cotton etc.) or in paper bags (e.g., maize) and applied to the receptive stigmas with the help of a camel hair brush, piece of paper, tooth pick or forceps. In some crops (e.g., Jowar, Bajra) the inflorescences of both the parents are enclosed in the same bag.

Harvesting and Storing the F_1 Seeds

Crossed heads or pods of desirable plants are harvested and after complete drying they are threshed. Seeds are stored properly with original tags.

Raising the F_1 Generation

In the coming season, the stored seeds are sown separately to raise the F_1 generation. The plants of F_1 generation are progenies of cross seeds and therefore are hybrids.

Hybridization Methods of Plant Breeding in Self-Pollinated Groups

Pedigree Method

Record of the ancestry of an individual selected plant for various generations is known as pedigree. A selection method, which is used in segregating population of self-pollinated species and keeps proper record of plants and progeny selected in each generation is known as pedigree breeding. This method is widely used for the development of varieties in self-pollinated crops.

In this method individual plants are selected till the progenies become homozygous. Selection for plants in the desired combination of characters is started in the F_2 generation and continued in succeeding generations until genetic purity is reached.

The method is as follows:

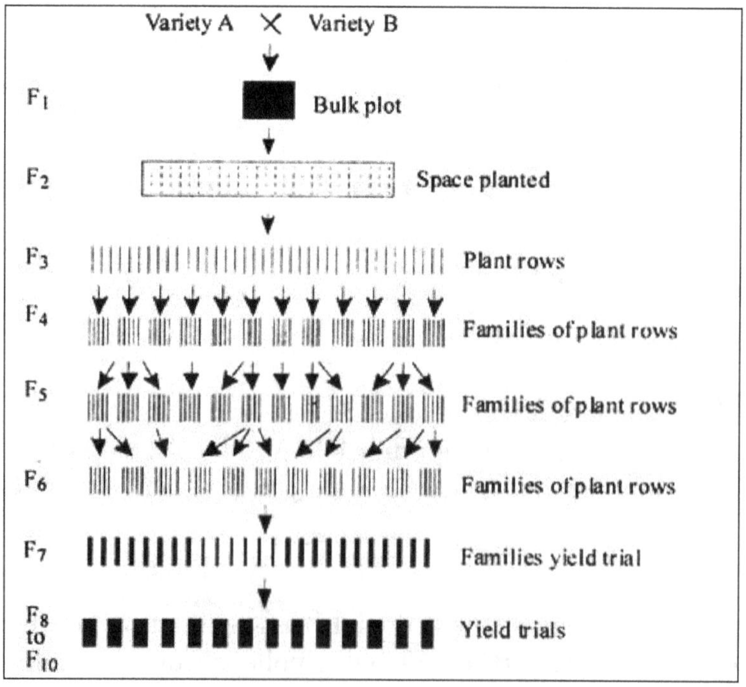

Different steps involved in pedigree method.

- I Year: Plants are chosen for hybridization and F_1 seeds are produced.
- II Year (F_1 generation): F_1 plants are space planted to produce maximum number of F_2 seeds.
- III Year (F_2 generation): 2000-10000 F_2 plants are space planted. About 200-500 desirable superior plants are selected.
- IV Year (F_3 generation): Selected superior plants in III year are space planted to study the individual plant. 3 to 5 best plants in these rows are selected and harvested (F_4).
- V Year and VI Year (F_4, F_5 generation): Process is continued as in F_3 generation. Normally 20-50 families may be retained at the end of F_5 generation.
- VII Year (F_6 generation): Due to successive self-pollination most of the lines become homozygous and uniform. The plants uniform in desired characters are harvested and the seed, bulked together to constitute the variety.
- VIII Year (F_7 generation): Preliminary yield trials are conducted.
- IX to XI year (F_8 - F_{10} generation): Trials of superior lines are confirmed. During the testing period observations are made on height, tendency to lodge, maturity, disease resistance and quality.
- XII to XIII Year (F_{10} - F_{11} generation): Seeds are multiplied and distributed to the farmers.

Merits

- It is the quickest method.
- Plant breeders can also obtain the genetic information.
- There are chances of recovering transgenic segregation by this method.

Demerits

- Maintenance of accurate pedigree record is not easy. It takes much time.
- Selected material becomes so large that handling of the same becomes very difficult.
- Success of this method depends upon the skill of the breeder.

Mass Pedigree Method

It is a modified form of pedigree method in which segregating material is handled by bulk (mass) method when conditions are unfavorable for selection and by pedigree method when conditions are favourable for selection.

Bulk Method or Breeding

A selection procedure which is used in segregating population of self-.pollinated species in which material is grown in bulk plot from F_2 to F_5 with or without selection, next generation is grown from bulk seed and individual plant selection is practiced in F_6 or later generations is called bulk method or breeding.

This method is also known as the mass or population method. Nilsson-Eule of Sweden was first to use the bulk method and it is in use ever since. This method differs from the pedigree method in that no selection is practiced in F_2 - F_5 generations.

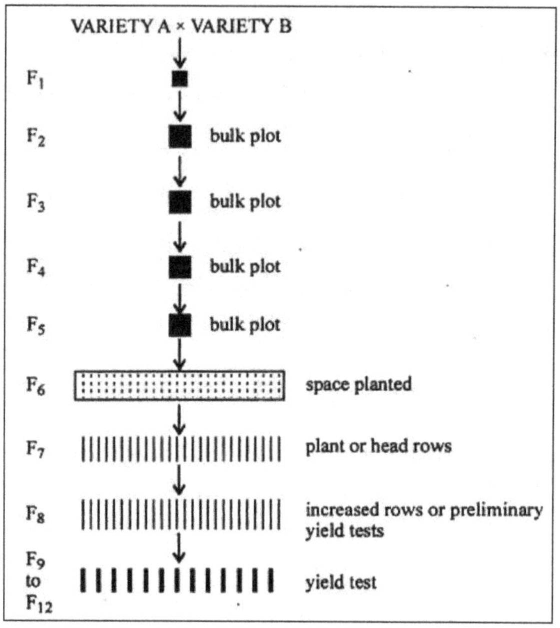

Procedure of bulk breeding method.

The method is as follows:

- I Year: Plants are chosen for hybridization and F_1 seeds are produced.
- II Year (F_1 generation): 50-100 F_1 plants are grown and their F_2 seeds are harvested in bulk,
- III Year (F_2 generation): F_2 plants are grown and their F_2 seeds are harvested in bulk.
- IV Year (F_3 generation): F_3 plants are grown and their F_4 seeds are harvested in bulk.
- V Year (F_4 generation): F_A plants are grown and their F_5 seeds are harvested in bulk.
- VI Year (F_5 generation): F_5 plants are grown and their F_6 seeds are harvested in bulk. (The process may be repeated until the desired period of homozygosity is achieved. In general bulk period is allowed up to F_5 generation).
- VII Year (F_6 generation): Seeds are space planted and single plant selection is done (F_7 generation).
- VIII Year (F_7 generation): The progeny of each single plant is grown separately and superior progeny are selected and isolated (F_8).
- IX Year (F_8 generation): Preliminary yield test are conducted (F_9).
- X-XII Year (F_9 - F_{12} generations): Multi-locations field trials are carried out, best performing strain is multiplied for seed distribution.

Merits

- The bulk method is simple, convenient, inexpensive and less labour consuming (no pedigree record is to be kept).
- During early segregating generations, very little work and attention is needed, which gives the breeder more time to concentrate on other breeding projects.
- Selection is done by nature only and it increases the frequency of superior types in the population.
- This method is suitable for studies on the survival of genes and genotypes in populations.

Demerits

- This method takes much longer time to develop a new variety.
- The breeder is enable to exercise his skill and judgement in selection and therefore the method is less satisfying to him.
- Information on the inheritance of characters cannot be obtained.
- This method is totally dependent on natural selection to select the superior types. These types may not be necessarily the best yielding types.

Single Seed Descent Method

This method was suggested by Coulden (1939) for advancing segregating generation of self-pollinated crops. A breeding procedure used with segregating populations of self-pollinated species in which plants are advanced by single seeds from one generation to the next is referred to as single seed descent method.

The procedure is as follows:

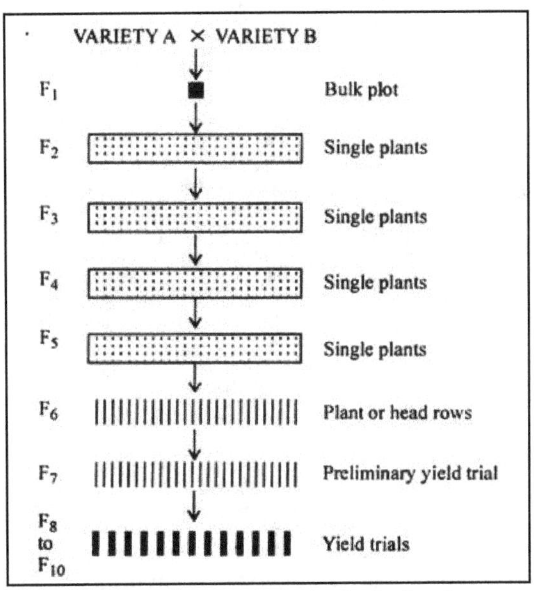

Single seed-descent method of selection.

Back Cross Method

This method was first proposed by Harlan and Pope (1922). Now-a-days this method is employed in improvement of both self and cross-pollinated crops where varieties are deficient in one or two aspects. This method is used particularly for transferring a single simply inherited character like disease, frost or drought resistance and earliness from an undesirable variety to a good commercial variety.

The desirable variety is called as recurrent or recipient parent and it is crossed to an undesirable variety, called as donor or non-recurring parent (called donor because the desirable genes are transferred).

F_1 plants instead of permitting to self-pollination as in pedigree or bulk method care crossed with the recurring parent and therefore, it is called as back cross method (A back cross may be defined as a crossing of F_1 hybrid with any of its parents).

The procedure is as follows:

Suppose there are two varieties A and B. A is very good in all characters but disease susceptible and the variety B is disease resistant but very poor in all characters. It is needed to transfer the disease resistance from B to A without adversely affecting the good qualities of A. B is then donor or non-recurring parent and A is recipient or recurring parent.

Selected plants of A and B are crossed to raise the F_1. Here A is treated as the female and B as male. F_1 plant is backcrossed with the recurrent parent; variety A and BC: plants (back cross first generation) plants are raised. Here B is treated as female and A as male. Select the BC_1 plants possessing desired characters of A with disease resistance of B.

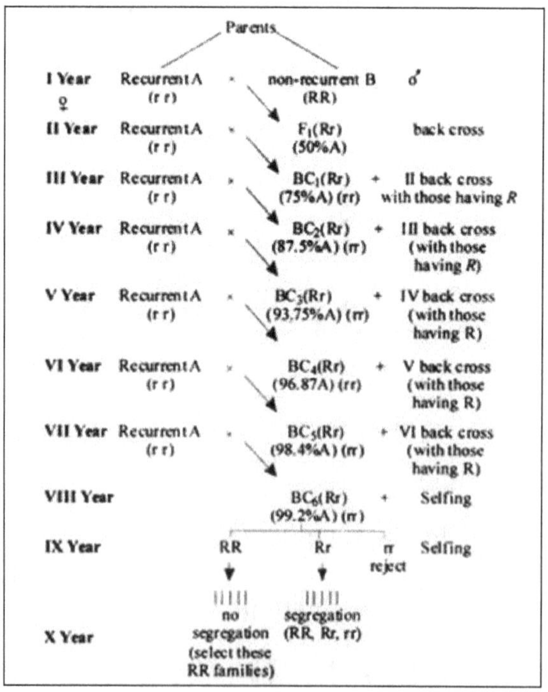

Different steps insolstd in the back cross method of plant breeding in a self pollinated crop. when a dominant trait is transferred from variety B to on otherwise insprosed variety A.

The selected BC_1 plants again backcrossed to A till a desirable type having good qualities of A and disease resistance of B is obtained. Disease resistant plants are selfed and individual plant progenies from the selfed seeds of selected plants are grown.

New variety is tested in replicated field trials along with the variety A as check. Seeds are multiplied and released for cultivation. The method is used for the transfer of a dominant gene. In case of recessive gene transfer, a slightly different procedure is followed.

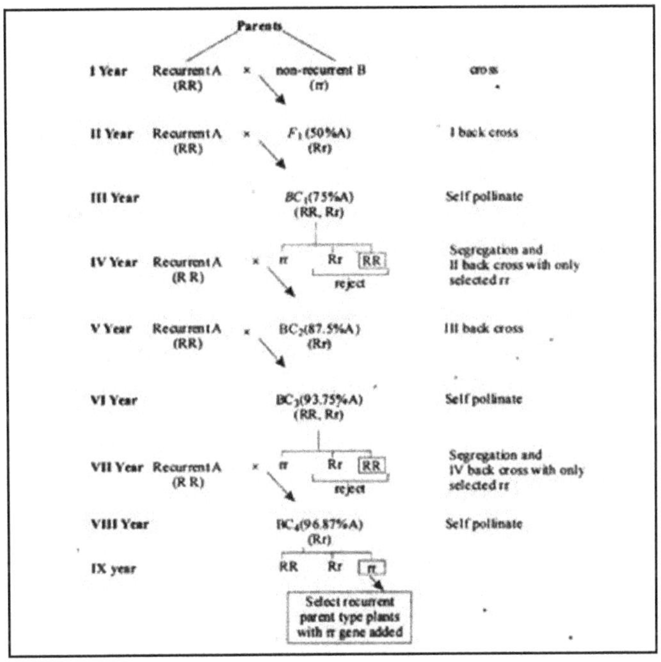

Different steps involved la the back cross method of plant breeding in a sell pollinated crop, when a recessive trait is transferred from a variety B to otherwise improved varlet A.

Merits

- This is the most common method used for transfer for oligogenic character like disease resistance.
- It is the only method used for the interspecific transfer of characters, transfer of qualitative characters, quantitative characters, transfer of cytoplasm, particularly for male sterility and for the production of isogenics lines.
- Role of environment is almost negligible. So time required for developing new variety may be drastically reduced because off-season nurseries and greenhouses can be used for developing 2-3 generations each year.
- Extensive yield tests are not required because the performance of the recurrent parent is already known.
- Breeder does not require to handle large populations.
- In this method all desirable characters of a popular variety (genotype known) are retained except the undesirable genes at a particular locus is replaced. So, the outcome of this method is known forehand, and plants can be reproduced any time in future.

Demerits

- The new variety is superior over the already available variety only in the transferred character.
- Hybridization has to be done for each backcross. It requires lot of time and money.
- Sometimes undesirable genes are closely linked with desirable one and get transferred to the new variety.

Multiple Cross Method

A cross-involving more than one inbred line is referred to as multiple cross. It is also known composite cross and is used to combine monogenetic characters from different sources into a single genotype. In this method, several pure lines are crossed together. The selected pure lines are first combined into crosses as A × B, C × D, E × F, G × H and so on.

The F_1 plants are mated together as (A × B) × (C × D) and (E × F) x (G × H). Finally, the F_1 plants of double crosses are crossed with each other to produce hybrids [(A × B) x (C x D)] x [(E x F) x (G x H)]. Further breeding in these hybrids is carried out according to either pedigree or bulk method.

A×B C×D E×F G×H

AB×CD EF×GH

ABCD × EFGH

ABCDEFG

Merits

- In self-pollinated crops this method is used when three or four monogenic characters scattered in three or four different varieties are to be combined into one.
- These crosses generally have wider adaptation.

Demerits

- These crosses are generally less productive.
- This method has limited utility except in high risk areas where severe disease damage occurs regularly from a highly specialized disease pathogen.

Self pollinated crops When the pollen of the same flower or other flowers of the same plant pollinates a receptive stigma, the condition so created is called self pollination.	
Annual sweet clover	Lespedeza species
Apricot	Lettuce
Barley	Lima beans
Bean	Linseed
Brinjal	Mountain bromegrass
Bur clover	Moong (Green gram)

Chick peas	Nectarine
Cotton	Oat
Cowpea	Pea
Foxtail millet	Peach
Groundnut or peanut	Pepper (Capsicum annum, C. fruiescens)
Lady's finger (Okra)	
Potato	Sweet pea
Rice	Tobacco
Soft chess	Tomato
Sorghum (h)	Urd
Soyabean	Velvet bean
Strawberry clover	Vetch
Subterranean clover	Wheat

Hybridization Methods of Plant Breeding in Cross-Pollinated Crops

Like self-pollinated crops, hybridization in cross-pollinated crops involves crossing of two or more inbreeds. The inbreeds are combined in any one of the following types of crosses and released as improved strains. These crosses are:

Single Cross

Crossing of two inbreeds or varieties are called single cross such as A x B or C x D. This cross was proposed by Shull (1909). The total number of possible different single crosses produced by a given number of inbreeds can be calculated by the formula n(n-1)/2 where n is the number of inbreeds. e.g., from four inbreeds six single crosses can be produced.

It can be calculated as:

$$\text{No of single crosses} = 4\frac{(4\text{-}1)}{2} = \frac{4\times 3}{2} = 6$$

Single crosses give the maximum degree of hybrid vigour but due to weak inbreeds, little amount of seed is produced.

Double Cross

In this method two single hybrids are crossed together to produce a double cross hybrid for e.g., if A, B, C and D represent inbred lines, one of the possible single cross can be represented by A x B and are of the possible double crosses by (A x B) x (C x D). The (n — 2) (n —3) number of double crosses can be calculated by the formula n $\frac{n(n-1)(n-2)(n-3)}{8}$ where n is the number of inbreds.

Thus with 4 inbreds 3 double crosses hybrids can be produced. It can be calculated as follows:

$$= \frac{4(4-1)(4-2)(4-3)}{8}$$

$$= \frac{4\times3\times2\times1}{8}$$

$= 3$ double crosses hybrids'.

These are:

1. $(A \times B) \times (C \times D)$
2. $(A \times C) \times (B \times D)$
3. $(A \times D) \times (B \times C)$

Commercial hybrids are produced by this method. These hybrids give very high yield in small land without any increase in the cost of production. All the hybrid seeds of maize to farmers for cultivation are nothing but double crosses.

Three-way Cross

It is a cross of a single cross (used as female) with another inbred (used as male), i.e., it involves three inbreeds [(A × B) × C], The single cross and the inbreeds are planted in the same way as the inbreeds in the single cross. The advantage of this cross is the use of vigorous hybrid of first generation as female in order to maximize the yield of hybrid seeds as well as to obtain seeds of normal grain size.

Top Cross or Inbred Variety Cross (A × variety)

It can be defined as a cross between an open pollinated variety and inbred line. Either the inbred or the variety may be used as female parent, but to use variety as female is preferable. It is used not only for the developing hybrid but to test the combining ability of the inbreeds.

Synthetic Cross

It can be defined as a cross among inbreeds, clones or sibbed lines without control of pollination. (Synonyms, poly-cross and strain building). This is the cross of a number (4-10) pretested hybrids and is done by open pollination in isolation.

Seeds of these inbreeds are mixed in equal proportion and sown in isolated plots. Natural cross pollination takes place and the harvested produce becomes synthetic cross. This cross is used in forage cross where floral structure causes difficulties in artificial pollination.

After any one of the above-mentioned crosses, seeds are tested, multiplied and distributed. The testing is done at various regional research stations by various research workers. The entire process takes about 5-7 years.

Cross pollinated crops When the pollen of a flower pollinates the stigma of another flower located on a different plant, it is termed as cross pollination	
Alfalfa or luceren	Hemp
Almond	Hops
Alsike clover	Kale
Apple	Kohlrabi
Artichoke	Maize
Asparagus	Mangel
Banana	Mango
Beet	Muskmelon
Birds	foot
trefoil	Olive
Blackberry	Onion
Blueberry	Orchard
grass	Broccoli
Broccoli	Parsley
Brussels sprout	Palestine (Strawberry clover)
Cabbage	Papaya (Papita)
Carrot	Pear
Castor	Pearl millet or bajra
Cauliflower	Pecan
Celery	Plum
Citrus	Pumpkin
Chard	Raspberry
Cherry	Red clover
Chicory	Rhubarb
Chinese cabbage	Rye
Chestnut	Safflower
Collard	Smooth bromegrass
Crimson clover	Spinach
Cucumber	Squash
Date palm	Strawberry
Fig	Sunflower
Filbert	Timothy
Graph	Turnip

Hybridization Methods of Plant Breeding in Vegetatively Propagated Crops

Improved clones of asexually propagating plants are selected and allowed to grow under conditions, which favour flowering and seed setting. Desirable clones are crossed together and the hybrids are multiplied by cloning. Each F_1 plant is a potential source for a new clone. This procedure has been used for developing many improved varieties of sugarcane, potato etc.

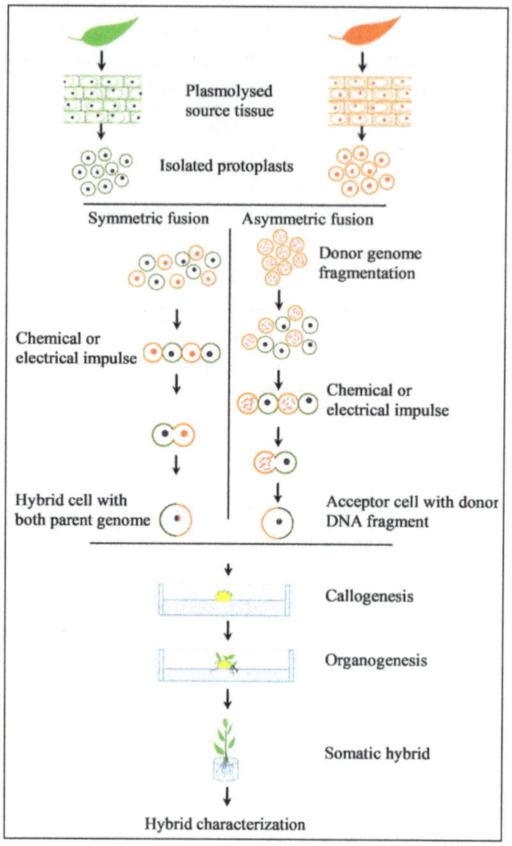

In Vitro Regeneration of Plants

The main techniques adopted for in vitro regeneration of plants are:

Micro-Propagation Technique

Micro propagation has been defined as 'in vitro regeneration of plants from organs, tissues, cells or protoplasts' and 'the true-to-type propagation of a selected genotype using in vitro culture techniques'. True-to-type propagation has important benefits for plants that have no named varieties, such as Australian dioecious papaw genotypes where traditional plant breeding has failed to produce stable lines.

Micro propagation has also been useful for the rapid initial release of new varieties prior to multiplication by conventional methods, e.g. pineapple and strawberry, and germplasm storage for maintenance of disease-free stock both in controlled environment conditions and long term via cryopreservation.

However, the ability to propagate plants in vitro free of genetic off- types is dependant on the technique used for micro propagation. Failure to understand this principle has resulted in disastrous consequences with some species, for example the commercial production of banana clones containing 90% dwarfs.

Protocols which have been developed for in vitro propagation of plant species can be divided into three systems. System 1 is based on callus culture followed by organogenesis or embryogenesis. System 2 comprises proliferation of axillary buds and/or adventitious buds, resulting from repeated subculture on multiplication media containing cytokinin.

In system 3, micro-cuttings obtained from axillary buds of apically dominant shoots are grown on hormone-free or low cytokinin media. System 3 has been used in the author's laboratory for clonal propagation of papaw, neem, coffee and passion-fruit and is not prone to production of genetic off-types.

Exploitation of the variation that exists in populations has led to the development of many commercial varieties and hybrids. With the rapid expansion of tissue culture technologies came the observation that genetic variation was occurring in plants regenerated from somatic cells and this was seen as a novel source of variation.

The frequency of off-types varies with species, culture type and number of sub-cultures and has been attributed to a number of variations within cultured cells. However, the cause of somaclonal variation is not fully understood; therefore, it cannot be controlled and changes can be epigenetic and unstable.

Variation can be promoted by the use of radiation or chemical mutagens and the use of colchicine to change ploidy. The potential of somaclonal variation involves the ability to change one or a few characters without altering the remaining part of a genotype.

Initial notes of potentially useful variation included increases in cane and sugar yield, and resistance to eye- spot disease in sugar cane; improved tuber shape, colour, and late blight resistance in potato; and, increased solids and Fusarium resistance in tomato.

In vitro selection involves the screening of cell cultures that are exhibiting genetic variation, for tolerance or resistance to pathogens, herbicides, low or high temperatures, metals and salt. It necessitates a reliable method of regeneration from callus.

For successful application of the technique, tolerance at the cellular level must also occur in the regenerated plant in the field and be transferable to other plants via conventional plant breeding.

The use of in vitro screening to select for disease resistance is most effective for diseases which produce toxins. This was initially demonstrated by Carlson who regenerated tobacco plants resistant to Pseudomonas tabaci, from haploid cells resistant to methionine sulfoxamine, an analogue of the disease toxin.

Similar successes have been noted with a range of diseases including Phoma lingam in canola, Helminthosporium oryzae in rice, Pseudomonas and Alternaria in tobacco and Pseudomonas solanacearum in tomato. Addition of herbicides to culture is an ideal system for in vitro screening as defined concentrations can be used.

However, effectiveness is dependent on the mode of action of a herbicide on the whole plant being similar at the cellular level, and is therefore better suited to herbicides that interfere with basic metabolic functions. Resistance or tolerance has been developed in vitro to a range of herbicides including paraquat in tomato, picloram amitrole and triazine in tobacco and 2, 4-D in Lotus corniculatus.

Carrot suspension cultures which were resistant to 35mM glyphosate in solution were selected after 8 subcultures on low concentrations of glyphosate (0.3 to 0.6 mM). Carrot plants regenerated from these cultures were resistant to glyphosate sprays in field plantings. A similar stepwise selection procedure for gene amplification produced tobacco cell lines with high tolerance to glufosinate.

Cell cultures that are resistant to increased salt (NaCl) in vitro do not necessarily produce regenerants which tolerate high salinity levels in soil. For example, the use of somatic cell cultures of rice did not lead to heritable salinity tolerance in the field; however, this was eventually achieved after in vitro selection of callus cultures initiated from immature ovaries and anther, culture-derived lines.

Examples of genotypes which had been released or were in field trials in the USA and Canada. Many of his references were patent applications or personal communications.

These included chlorsulfuron and imidazilinone resistance in canola, imidazilinone resistance in corn, high solids and Fusarium race 2 resistance in tomato, a white flowered form of lucerne, potato virus Y resistance in tobacco and fall armyworm resistance in sorghum.

Potentially useful somaclonal variants have been identified in a number of fruit crops. These include Phytophthora resistance in papaya and apple, Xanthomonas resistance in peach, Erwinia resistance in pear and salt tolerance and 2, 4-D tolerance in orange.

Embryo Culture Technique

Embryo culture was probably the first tissue culture technique to be applied to plant improvement. Applications of embryo culture are rescuing embryos after interspecific hybridisation, clonal propagation of families such as Gramineae and conifers which contain recalcitrant species, and overcoming seed dormancy and seed sterility.

Interspecific crosses are often attempted to transfer desired traits such as disease resistance, stress tolerance or high yield from wild species into important crop species, e.g. cotton, soybean and papaya.

Incompatibility after these crosses normally results in embryo abortion and this is often caused by breakdown of the endosperm or embryo- endosperm incompatibility. In vitro culture of hybrid embryos often successfully bypasses post zygotic incompatibilities.

There are numerous references of embryo rescue of incompatible hybrids. Another approach to wide hybridisation is in vitro fertilisation of cultured ovules and ovaries. This was achieved initially with opium poppy. The major application of this technology in monocotyledonous crop species is in maize.

In an extensive research program funded by the Australian Centre for International Agricultural Research (ACIAR) techniques have been developed for interspecific hybridisation and embryo rescue of papaya and wild relatives.

Hybrids have been produced between C. papaya and C. cauliflora, C. quercifolia, C. pubescens, C. parviflora and C. goudotiana. Useful characteristics of the wild species include papaya ring spot virus resistance, Phytophthora resistance, high sugar content, and ornamental characteristics.

Gene Transfer Techniques

Many years of research have culminated in an increasingly detailed understanding of biology and genetics at the molecular level. Combined with tissue culture technology, this knowledge is being applied to the modification of plant genomes. Because of the universality of the genetic code, potentially useful genes can be transferred to plants from any organism.

This has lead to the development of new and improved genotypes by the addition of single genes that code for traits such as insect or disease resistance, or by the inactivation of single gene faults. Future potential includes the possibility of transferring larger DNA constructs that code for multiple genes conferring more complex traits.

Applications of gene transfer are increasing rapidly, driven by an extensive research effort worldwide. Initial successes were in the development of disease, insect and herbicide resistant crops.

More recently, gene transfer is resulting in control of plant development in individual species. Inhibition of expression of polygalacturonase in a tomato genotype lead to the first commercial release of a genetically engineered plant – the FLAVR-SAVR tomato. Extensive research is being directed towards gene inactivation techniques for control of ethylene production, and thus ripening, in climacteric fruits.

Gene transfer techniques have also lead to the development of carnations with increased vase life and new colour forms. Isolation of a flower-meristem-identity gene from Arabidopsis has the potential to dramatically reduce time to flowering and hence length of the juvenile phase in plant species.

Developing applications include transformed plants which will produce more or increased levels of valuable oils, proteins and starches, and plants that produce vaccines and can be used for oral immunisation against a range of serious human diseases.

Useful agricultural applications of gene transfer necessitates incorporation of a foreign DNA construct into a plant genome, the regeneration of transformed plants, the stable expression of the introduced gene and inheritance in subsequent generations.

The successful application of plant transformation techniques is dependent on plant tissue culture protocols to regenerate transformed plants. Since the first reports of gene transfer with species that are relatively easy to tissue culture, petunia and tobacco, transformation procedures have been published for a wide range of species including tree crops, such as apples and papaws.

Transgenic plants of a number of fruit species have been produced, including kiwi-fruit, citrus, strawberry, grape, cranberry, peach and plum. Numerous DNA delivery systems have been noted.

Direct gene transfer systems such as microinjection, electroporation and polyethylene glycol have been used to transform protoplasts. Protoplasts are easily damaged and regeneration of plants from them is very difficult with many species.

Consequently, such systems are not widely used. The popular method of direct gene transfer is particle bombardment. DNA coated micro particles are bombarded into tissues, and major crop species have been transformed using this technique, for example monocots such as wheat, barley, rice and recalcitrant dicot species such as cotton, and soybean.

The advantages of the particle gun in terms of tissue culture regeneration is that it can be used on callus and suspension cultures and on organised tissue of both monocots and dicots.

Increasing the number of options for tissue culture technology increases the probability of successful regeneration. The preferred method of gene transfer for dicots is Agrobacterium mediated transfer.

Co-cultivation of callus, suspension cultures or leaf discs with Agrobacterium has been used to successfully transform many species, for example efficient systems have been described for major crop species such as potato, tomato and sugar beet.

Virus, insect and herbicide resistance in a range of species including tobacco, potato, tomato, lucerne and soybean have been successfully trialed in field conditions and are in the early stages of commercialisation. Transformation of cereals, which are the world's most important food, has been by direct DNA uptake systems, particularly biolistics.

Embryo genic cultures in cereals have generally been developed from immature embryos and inflorescences and subsequently transformed in the form of protoplast, suspension or callus cultures. The most recalcitrant of these species is wheat, where difficulty in developing reliable regeneration systems is exacerbated by low frequency of transformation.

Other problems associated with tissue culture of cereals are loss of regenerative capacity with increasing number of subcultures and occurrence of abnormalities such as sterility in regenerated plants. Agrobacterium was thought incapable of infecting monocots, however recent notes describe Agrobacterium mediated transformation of rice, maize and barley which offer considerable potential.

Systems have been noted for transformation of haploids via particle bombardment of pollen and via microinjection of DNA into microspore derived embryoids of canola. More recently, maize has been transformed by pollen which was mixed with Agrobacterium containing binary plasmids.

Such systems may provide a method that eliminates the need for in vitro regeneration, which is a major limiting step for some species and is prone to somaclonal variation.

Although in theory tissue of any species can be transformed, it is currently not possible to regenerate transformed plants at high frequencies for many species. Current methods of gene transfer and selection can be disruptive to growth of in vitro cultures, thus it can be difficult to develop efficient transformation procedures for species that are recalcitrant in vitro.

In addition, there is a need to be able to deliver gene constructs to tissue that is competent both for transformation and regeneration.

Limitations of current technology include low frequency of transformation, high frequency of undesired genetic aberrant, unpredictability of transgene expression, the need for techniques to transfer large DNA sequences coding for multiple genes and the absence of efficient repeatable regeneration protocols for many species in vitro.

In Vitro Techniques

In Vitro Collecting

In vitro collecting involves initial disinfestation and placement of plant explants in sterile culture medium, before transport to a tissue culture laboratory for further in vitro procedures. In vitro collection is particularly useful for species that are vegetatively propagated and for those with recalcitrant seeds or embryos which deteriorate rapidly.

The technique has much potential to facilitate the collection of germplasm of tropical and subtropical fruit species, as has already been demonstrated with Cassava and coconut. Recently, 300 Musa accessions were collected in Papua New Guinea using this technique, before being transported to a collection in Australia.

An added advantage of this exercise is that it complied with quarantine regulations that are in place to stop the spread of Fusarium and other diseases.

In Vitro Culture

In vitro culture offers some major advantages for conservation and use of plant genetic resources. In vitro cultures can be established from disease-free parent material and maintained in a disease-free state. Alternatively, meristem culture, in combination with treatments such as chemo- or thermotherapy, is a proven technique for elimination of specific viral and other diseases.

Consequently, disease-free germplasm can be safely and rapidly transferred between countries. However, it is not safe to assume in vitro cultures are free of specific viruses, unless the initial parent material has been cleared by appropriate ELISA or equally stringent testing. Caution should always be exercised when dealing with export/ import of plant tissue cultures.

In vitro culture has been used to facilitate international collaboration in a research project between Australia and the Philippines on development of PRSV-P resistance in papaya. Sterile cultures of Carica species, interspecific hybrids and embryo genic cultures were regularly transferred between the two countries and contained in quarantine approved laboratories.

For example, cultures from the Philippines have been transported to Australia, subjected to in vitro procedures in the author's laboratory and returned to the Philippines. Australian quarantine restrictions that prevent early field release of this material are not contravened.

In Vitro Germplasm Storage

Ninety percent of all accessions held in germplasm collections are stored as seeds. There are three limitations of seed storage. Firstly, it is not applicable to species with no seeds e.g. Musa species and Artocarpus altilis (breadfruit). Secondly, for highly heterozygous species, e.g. dioecious papaya, it is preferable to conserve vegetative or clonal material.

Thirdly, seeds of recalcitrant species are sensitive to desiccation and/or chilling, and tropical recalcitrant species often lack a natural dormancy mechanism, e.g. Mangifera indica (mango), Nephelium lappaceum (rambutan) and Cocus nucifera (coconut).

In vitro storage also represents a viable alternative to field collections for species included in the three groups for rare and endangered species. Field collections require regular maintenance and are prone to damage from disease and insect attacks, extremes of weather and natural disasters.

By comparison in vitro collections are safe from these problems, although failures in environmental control can result in loss of cultures. Thus duplicate collections are advisable. Another advantage of in vitro storage over field collections is that large numbers of cultures can be stored in a small volume in a growth room.

In vitro collections are generally maintained in either slow growth storage or by cryopreservation. Slow growth storage employs various methods to reduce sub-culture frequency, and thus labour and media costs. Cryopreservation involves long term growth suspension in liquid nitrogen. Cryopreservation was achieved initially using cryoprotectants and controlled cooling.

This classic technique has been successfully applied to cell suspension and callus cultures. Newer methods involve dehydration of cultures before rapid freezing by immersion in liquid nitrogen. This method shows potential with organised tissue, including embryos and meristems.

There exist a number of different dehydration/freezing procedures, including vitrification, encapsulation dehydration, desiccation, pre-growth and droplet freezing.

References

- Reproduction-in-Plants: biologyreference.com, Retrieved 30 August, 2019
- Asexual-reproduction-plants, biology: byjus.com, Retrieved 16 January, 2019
- Sexual-reproduction-plants, biology: byjus.com, Retrieved 29 March, 2019
- Sexual-reproduction-in-plant, reproduction-in-plants, science: toppr.com, Retrieved 20 April, 2019
- Pollination, sexual-reproductio-in-flowering-plants, biology: toppr.com, Retrieved 2 June, 2019
- Sexual-reproduction-in-plants, biology: lumenlearning.com 22 May, 2019
- Pollination-Techniques: plantbreeding.coe.uga.edu, Retrieved 23 February, 2019
- Sexual-reproduction-in-plant, reproduction-in-plants, science: toppr.com, Retrieved 10 July, 2019
- Double-fertilization-in-angiosperms, biology: byjus.com, Retrieved 13 March, 2019
- Gametogenesis-in-plants, sexual-reproduction-in-flowering-plants, biology: toppr.com, Retrieved 29 June, 2019
- Hybridization-method-of-crop-improvement, crops: biologydiscussion.com, Retrieved 31, March, 2019
- In-vitro-regeneration-of-plants, techniques-biotechnology, genetically-modified, biotechnology: biotechnologynotes.com, Retrieved 14 July, 2019

3

Selection Methods for Plant Breeding

Natural selection and artificial selection are the two broad classifications of plant breeding. Natural selection method involves the natural processes and functions in nature while human beings are the selecting agents in artificial selection methods. All the diverse selection methods for plant breeding have been carefully analyzed in this chapter.

Natural Selection

This is a natural process. It operates in the nature without human interference. According to the Darwin's principle "Survival of the fittest" plants which survive through the adversities of nature are preferred and the weaker ones are wiped out. Thus, nature itself selects the fittest organisms.

So, natural selection favours these characters which are essential for survival of a species. The selection pressure ultimately resulted in the appearance of many differences between species and subspecies. Natural selection has given the cultivated crops and 'ecotypes' in plants.

Natural selection can take many forms. To make talking about this easier, we will consider the distribution of traits across a population in graphical form. For example, if we were talking about height as a trait, we would see that without any selection pressure on this trait, the heights of individuals in a population would vary, with most individuals being of an average height and fewer being extremely short or extremely tall. However, when selection pressures act on a trait, this distribution can be altered.

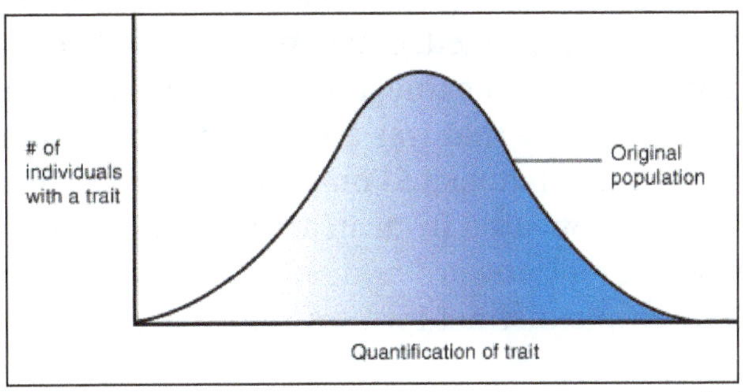

Figure: The standard distribution of a trait across a population.

Stabilizing Selection

When selective pressures select against the two extremes of a trait, the population experiences stabilizing selection. For example, plant height might be acted on by stabilizing selection. A plant that is too short may not be able to compete with other plants for sunlight. However, extremely tall plants may be more susceptible to wind damage. Combined, these two selection pressures select to maintain plants of medium height. The number of plants of medium height will increase while the numbers of short and tall plants will decrease.

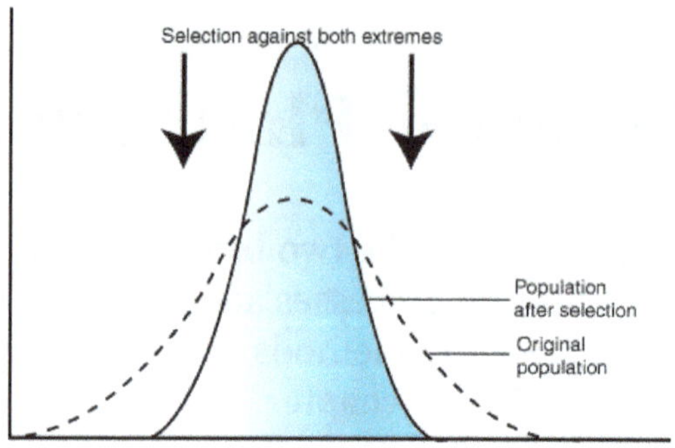

Figure: The effect of stabilizing selection on trait distribution.

Directional Selection

In directional selection, one extreme of the trait distribution experiences selection against it. The result is that the population's trait distribution shifts toward the other extreme. In the case of such selection, the mean of the population graph shifts with long necks.

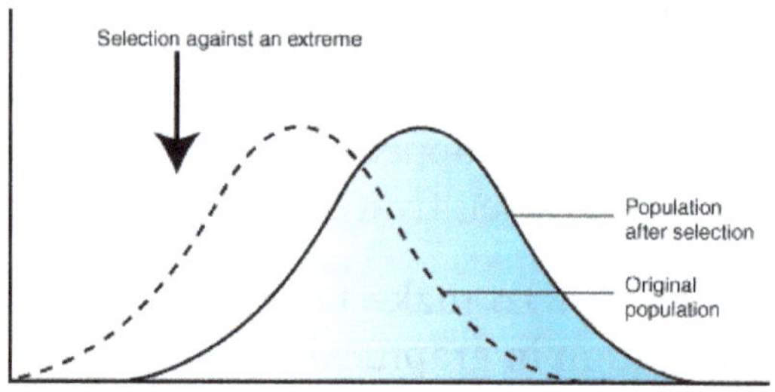

Figure: The effect of directional selection on trait distribution.

Disruptive Selection

In disruptive selection, selection pressures act against individuals in the middle of the trait distribution. The result is a bimodal, or two-peaked, curve in which the two extremes of the curve create their own smaller curves. For example, imagine a plant of extremely variable height that is pollinated by three different pollinators, one that was attracted to short plants, another that preferred plants of medium height and a third that visited only the tallest plants. If the pollinator that preferred plants of medium height disappeared from an area, medium height plants would be selected against and the population would tend toward both short and tall, but not medium height plants. Such a population, in which multiple distinct forms or morphs exist is said to be polymorphic.

Artificial Selection

In artificial selection, certain individual plants are chosen for the purpose of having better crop from a mixed population where the individuals differ in characters. Here the selecting agent is man. Man exploits the variations existing among the species. He picks of a few plants of better qualities from mixed populations and tries to propagate them.

Mass Selection

Mass selection is one of the oldest methods of crop improvement. In this method, individual plants are selected on the basis of phenotype from a mixed population, their seeds are bulked and used to grow the next generation. In other words, it is a method of breeding in which individual plants are selected on the basis of phenotype from a mixed population, their seeds are bulked and used to grow the next generation.

The main features of varieties developed by mass selection in self and cross pollinated species are given below:

Application

Mass selection is applicable to both self and cross pollinated species. However, it is more commonly used in the improvement of cross pollinated crops than in self pollinated species. This method is rarely used in vegetatively propagated crops.

Genetic Constitution

In self pollinated crops, a mass selected variety is homozygous but heterogeneous, because it is a mixture of several pure-lines. In cross pollinated crops, such varieties are mixture of homo and heterozygotes and are heterogeneous, because they consist of several homo and heterozygous genotypes.

Adaptation

Mass selected varieties have wide adaptation and are more stable against environmental changes due to heterogeneity which provides better buffering capacity. In other words, mass selected varieties have broader genetic base than pure lines. They exhibit more or less stable performance. However, adaptability is more in cross pollinated crops than in self pollinated species.

Variation

They are composed of several pure lines in self pollinated crops and of several homo and heterozygous genotypes in cross pollinated crops. Hence there is heritable variation in the mass selected varieties, besides environmental variation. The heritable variation provides them good buffering capacity.

Selection

Selection is effective in case of mass selected varieties of self pollinated crops due to presence of heritable varieties. However, further selection in the mass selected varieties of cross pollinated crops may lead to inbreeding depression.

Quality

A variety developed by mass selection is less uniform in the quality of seed than pure-lines due to presence of heritable variation.

Resistance

Mass selected varieties are less prone to the attack of new diseases due to genetic diversity. In other words, they are more resistant or tolerant to new diseases.

Roguing

Periodic removal of off type plants is essential to maintain the yield of mass selected varieties.

Types of Mass Selection

There are two types of mass selection, viz.:

1. Positive mass selection
2. Negative mass selection

Positive Mass Selection

When desirable plants are selected from a mixed population and their seeds are mixed together to grow further generation, it is referred to as positive mass selection. This process is continued for several years. Generally, old varieties or land races are used as the base material for mass selection. Selection of desirable plants or positive approach is in common use in mass selection.

Negative Mass Selection

When only undesirable off type of plants are removed from the field and rest are allowed to grow further, it is known as negative mass selection. This is generally used for varietal purification in seed production and certification programs. This helps in maintaining high level of genetic purity in the varieties especially in the self pollinated species.

The success of mass selection mainly depends on three factors, viz.:

1. Variability in the base population,
2. Mode of inheritance of character to be improved,
3. Heritability of the character.

Mass selection is more successful in old heterogenous variety or land races than in improved varieties.

Modifications of Mass Selection

Mass selection is practiced both in self and cross pollinated crop plants. However, it is more common in cross pollinated species than in self pollinated species.

There are two defects of mass selection as given below:

1. No Control on Pollination: In mass selection there is no control on the pollination. The selected plants are pollinated both by superior and inferior pollen parents.

2. Selection is based on Phenotype: The phenotypic performance is greatly influenced by environmental factors such as soil heterogeneity.

In order to overcome these defects, three modifications of mass selection have been suggested in cross pollinated species.

These modifications are:

1. Rejection of inferior pollen plants,
2. Use of composite pollen,
3. Stratification of the field.

A brief description of each method is presented below:

1. Rejection of Inferior Plants: In this method inferior pollen parents are removed before flowering and rest are allowed for inter-mating. This controls pollination by inferior plants.
2. Use of Composite Pollen**:** In this method, pollen is collected from the selected plants and bulked. This composite pollen is used to pollinate the selected plants. This also controls pollination by inferior plants.
3. Stratification of Field**:** This is known as grid method of mass selection and was suggested by Gardner. In this method, the field in which mass selection is to be practiced is divided into small plots in such a way that each plot should have 40-50 plants. Now superior plants are selected in each small plot. This eliminates the effect of environment [soil heterogeneity] because small plots are more homogeneous than larger plots.

The success of mass selection is more with oligogenic recessive characters than with polygenie characters. In case of polygenic traits, large number of individual plants should be selected in each generation to tap maximum desirable alleles. Moreover, mass selection is more effective for characters having high heritability. Generally, oligogenic characters have higher heritability than polygenic characters.

In cross pollinated species, the progeny of each cycle of mass selection should be grown in isolation to prevent contamination with other varieties (mating types) of a crop. The isolation distance differs from crop to crop depending upon the agency of pollination.

Generally, more isolation distance is required for wind pollinated species than insect pollinated ones. In general, isolation distance of 300 metres in case of insect pollinated species and 500-1000 metres for wind pollinated species is sufficient. In self pollinated species distance of 5-10 metres is adequate.

Procedure of Mass Selection

Mass selection consists of various steps, viz., selection of base population, selection of desirable plants from base population and mixing their seeds to rise next generation, evaluation in field trials, and releasing as a new variety.

The general procedure of mass selection is given below:

- First Year: An unimproved old variety or land race is used as a base population which is grown in a large plot. Then individual plants (55-1000) are selected on the basis of phenotypic performance for characters like height, maturity, disease resistance, productivity etc. The selected plants are harvested at maturity and their seeds are mixed together to grow next generation. This process is repeated till desirable results are achieved.
- Second Year: The crop is grown from the bulk seed of selected plants in a separate field using standard variety as a check for comparison of performance. In other words, the material is evaluated in preliminary yield trial. If mass selection is used for purification of old mixed variety, the same old variety can be used as check for comparison.
- Third to Sixth Year: The performance of bulk is evaluated for yield and adaptation in main yield trials for 3 to 4 years using standard check for comparison.
- Seventh and Eighth Year: The variety is released and named in seventh year and seed is multiplied. In the eighth year the seed is ready for distribution.

Merits of Mass Selection

- This is a good method for improvement of old varieties and land races. This is also used for the purification of improved varieties.
- Mass selected varieties are more stable in their performance than pure-lines. In other words, they have more buffering capacity than pure-lines due to heterogeneity.
- Mass selected varieties provide good protection against diseases.
- Mass selection is a simple and quick method of crop improvement. It takes about 8 years for the release of a new variety, whereas pure-line selection takes about 10 years in the development of new variety.
- This method is applicable to both self and cross pollinated species.

Mass selection has played more significant role in developing new varieties in cross pollinated species than in self pollinated species.

In India, mass selection has been useful in the development of improved varieties in cross pollinated crops like maize, pearl millet and mustard, and in often cross pollinated species like cotton and sorghum. In self pollinated species it has been rarely used because pure line selection is more effective in such crops than mass selection.

Demerits of Mass Selection

- The selection is based on the phenotypic performance. The superior phenotype is not always an indication of superior genotype. The real breeding value of single plants can be judged from the performance of their progeny. Progeny test is not carried out in mass selection.
- In cross pollinated species, there is no control on the pollination. The selected plants are

pollinated by both superior and inferior pollen parents. This results in rapid deterioration of variety developed by mass selection.

- In cross pollinated crops, large numbers of plants have to be selected for bulking, because small sample will lead to inbreeding depression.
- The produce of varieties developed by mass selection is less uniform man pure lines. This is because mass selected varieties are mixture of several pure lines in self pollinated crops and consists of several genotypes in cross pollinated species.
- In self pollinated species, pure line selection is more effective than mass selection, because pure line selection leads to isolation of best line from a mixed or heterogeneous population.

Pure-Line Selection

In Pure Line Selection a large number of plants are selected from a population of self-pollinated crop, harvested individually and their individual progenies are evaluated and the best progeny is released as a pure line variety. So the pure line is a progeny of a single homozygous self-pollinated plant.

Purpose of Pure-Line Selection

In self-pollinated crops pure-line selection method has several applications:

- Improvement of Local Varieties: It is more favourite method of improvement of local varieties which have considerable genetic variability.
- Pure Line Selection for Introduced Varieties: Introduced varieties are often subjected to pure line selection to develop suitable varieties for that place.
- Improvement of Old Pure Line Varieties: Pure lines become variable with time, i.e., off type plants appear in pure line. From such genetically variable pure lines often again pure line selection method is applied to make it more stable.
- Selections for New Characteristics in a Pure Line: Sometimes new favourable characters which were not considered previously can be selected from the existing variety through pure line selection method.
- Pure Line Selection in Hybridization Programme: Pure line selection method is most applicable in hybridization programme, i.e., selection from segregating generations of crosses. The pure lines in such cases may be selected using pedigree, bulk or backcross methods.

Procedure

The general procedure of pure line selection involves three steps:

- Selection of individual plants is done from a local variety. The superior and similar observations are selected assuming those are homozygous.
- Visual evaluation of individual plant progenies is performed and the superior progenies are harvested separately.

- Replicated yield trials are carried out for critical evaluation of the progenies. The most promising strain is identified and released as variety.

The procedure for pure line selection may be outlined in brief as follows:

- First Year: A large number of plants (200-3000) are selected from 'desi' or local variety or any other mixed population. The seeds are harvested from individual plant separately. The selection is made on easily observable characters like flowering, maturity times, disease resistance, presence of awns, plant height, etc.
- Second Year: Progenies from individual selected plants are grown separately with proper spacing. The progenies are observed visually and poor, weak or defective progenies are rejected, selection is made for very much desirable characters. Disease epiphytotics may be created to test the progenies for disease resistance. If necessary this process may be repeated for one or more years.
- Third Year: This step consists of replicated yield trials for critical evaluation of the progenies. The selected progenies are planted in replicated trial using the best variety as check. If enough seeds are available then preliminary yield trial may be conducted.
- Fourth Year: Replicated yield trials are conducted by the breeder using the best variety as check. The promising strains are observed and harvested for coordinated yield trial for further evaluation.
- Fifth to Eighth Year: The promising strains are evaluated at several locations along with other strains. The most promising strains are identified and released as new varieties.
- Ninth Year: The best progeny seeds are multiplied and released as a new variety through seed certification.

First Year	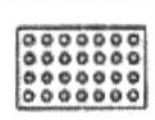 Desi or Old Variety (A mixture of pure lines)	(a) 200-3000 plants are selected on the basis of their perior and similar phenotype.
Second Year	 Individual Plant Progenies	(b) Individual plant progenies are grown separately. (c) Undesirable progenies are rejected.
Third Year	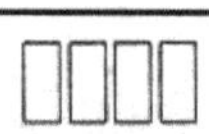 Preliminary Yield Trial	(d) Remaining progenies are planted for preliminary yield trial. (e) Low yield or inferior progenies are rejected.
Fourth to Seventh Year	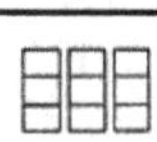 Multilocation Yield Trials	(f) Replicated yield trials conducted at several locations. (g) Inferior progenies are rejected. (h) Disease resistance and quality tests are done.
Eighth Year	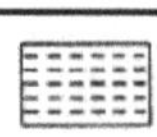 Seed Multiplication	(i) Best progency released as a new variety. (j) Seed multiplication and distribution are performed.

The total time required for the whole programme may vary depending upon several factors. If enough seeds are available in the second year then preliminary yield trial is done in the third year. Depending on the performance in the coordinated multi-location yield trial the time period can be lessened.

Merits of Pure-Line Selection

1. Pure line selection can achieve maximum possible improvement over the original variety. This is because the variety is the best pure line present in the population.
2. The pure line variety is extremely uniform since all the plants in the variety have the same genotype. Such uniformity is liked both by the farmers and the consumers.
3. Due to extreme uniformity, the variety is easily identified in seed certification programme.

Demerits of Pure-Line Selection

1. As the pure line variety is genetically more homogeneous, the adaptability to various environments is less.
2. This procedure requires more time, space, labour and more expensive yield trials.
3. The upper limit of improvement is set by the genetic variation present in the original population.
4. Once a pure line is developed, the variability is frozen and further selection becomes less effective.

Progeny Selection

Progeny selection is defined as follows:

1. A selection procedure in which superior plants are selected from a heterogeneous population on the basis of the performance of their progeny is referred to as progeny selection.
2. Selection of plants from a diverse population on the basis of their progeny test is called progeny selection.

The genetic worth of an individual is assessed by progeny test. The test of the genotypic value of an individual based on the performance of its progeny is called progeny test. The progeny test was developed by Louis de Vilmorin, hence it is also known as Vilmorin principle.

The progeny test is useful in two ways, viz.:

1. In understanding whether a plant is true breeding (homozygous) or segregating (heterozygous) for a particular character.
2. In the assessment of the breeding value of a plant. If the selected plant is really superior, its progeny will also exhibit superior performance and vice versa.

Features of Progeny Selection

1. Application: Progeny selection is commonly used in cross pollinated and often cross pollinated species.
2. Base Material: In cross pollinated species, three types of materials, viz.:
    - Open pollinated seeds,

- Self seeds,
- Top cross or test cross seeds of selected plants can be used in producing the progenies for testing. Generally, 10-50 seeds of each selected plant are grown for progeny testing.

3. Basis of Selection: In this method, selection of superior plants is based on the progeny performance (genotype). Those plants whose progeny performance is superior for the character under consideration are bulked together to produce the next generation.
4. Conduct of progeny test: Progeny test should be conducted in replicated trial to get more reliable results. The multi-locational or multi-seasonal test is considered ideal for progeny testing. Finally, plants whose progeny performance is good are selected and rests are discarded.
5. Genetic constitution: A variety developed by progeny selection is heterogeneous, because it consists of several homo and heterozygous genotypes.
6. Adaptation: A variety developed by progeny selection has wide adaptation and is more stable against environmental changes due to heterogeneity which provides better buffering capacity.
7. Variation: They are composed of several homo and heterozygous genotypes. Hence there is heritable variation in the varieties developed by progeny selection. The heritable variation provides them good buffering capacity.
8. Selection: Further selection in a variety developed by progeny selection may lead to inbreeding depression.
9. Quality: A variety developed by progeny selection is less uniform in the quality of seed than pure lines due to p resence of heritable variation.
10. Resistance: A variety developed by progeny selection is less prone to the attack of new diseases due to genetic diversity.

Example: The ear to row selection which is used in maize is a simple form of progeny selection.

Comparison of Mass Selection and Progeny Selection

S. No	Particulars	Mass selection	Progeny Selection
1	Applicable to	Both self and cross pollinated crops	Cross and often cross pollinated crops
2	Selection is based on	Phenotypic performance	Progeny performance
3	Adaptation of variety	Wide	Wide
4	Control over pollination	No control	No control
5	Genetic Composition	Heterogeneous	Heterogeneous
6	Selection	Effective after 4-5 generations	Effective after 4-5 generations
7	Time taken to release new variety	8-10 years	8-10 years
8	Vulnerability to new race of a disease	Low	Low
9	Exact reconstitution	Not possible	Not possible
10	Maintained by	Open pollination	Open pollination

Merits of Progeny Selection

- Selection is based on progeny performance which is a reliable technique.
- This method is very simple and convenient.

Demerits of Progeny Selection

- The main demerit of progeny selection is that there is no control on the pollination. The plants of superior progeny are also pollinated by the plants of inferior progenies.
- If the progeny of each plant is tested in isolation, it would require lot of area which is not practically possible.

Clonal Selection

A clone is group of plants produced from a single through asexual reproduction. Thus asexually propagated crops consist of large number of clones, and they are often known as clonal crops. All the members of a clone have the same genotype as the parent plant. As a result, they are identical with each other in genotype. Consequently, the phenotypic differences within a clone do not have a genetic basis and are purely due to the environmental effects. A selection within a clone is thus useless.

Identical Genotype

All the individuals belongings to a single clone are identical in genotype. This is so because a clone is obtained through asexual reproduction, which involves mitotic cell division only. Genetic variation in the progeny of a plant is produced chiefly by segregation and recombination, which occur during meiosis only. Thus the genotype of a clone is maintained indefinitely without any change.

Lack of Genetic Variation

The.phenotypic variation present within a clone is due to the environment only.This is so because all the individuals belonging to a single clone have the same genotype.

The phenotype of a clone is due to the effects of genotype (G), the environment (E) and the genotype X 'environment interaction (G x E) the population mean (μ). Thus the phenotype (P) of a clone may be expressed as follows: $P = \mu + G + E + GE$.

Thus the phenotypic differences among clones would be partly due to E and GE components. Hence the efficiency of selection among clones, as among purelines, would depend upon the precision with which the E and GE components of phenotype are estimated.

Immortality

Theoretically, clones are immortal i.e., a clone can be maintained indefinitely through asexual reproduction. But clones usually degenerate due to viral or bacterial infection. A clone may become extinct due to its susceptibility to diseases or insect pests. Further, genetic variation may arise within a clone changing its characteristics.

Severe Inbreeding Depression

Generally, clones are highly heterozygous and show severe loss in vigor due to inbreeding.

Clonal Selection Process

The phenotypic value of a plant or clone is due to the effects of its genotype (G), the environment (E) and genotype x environment (G x E) interaction. Of these, only the G effects are heritable. The environmental and interaction effects are non heritable and cannot be selected for. Therefore, a selection for quantitative characters based on observations on single plants is highly unreliable. In fact, plants selected in this way may be no better than a random sample.

Further, a selection for characters like yielding ability, etc. on the basis of unreplicated clonal plots would often be misleading and unreliable. Therefore, the value of a clone can be reliably estimated only through replicated yield trials. However, selection for highly heritable characteristics, such as plant height, days to flowering, color, disease resistance, etc., are easy and effective even on the basis of individual plants or single plots. Clearly, these situations are the same as those in the case of sexually reproducing crops.

In view of these considerations, in the earlier stages of clonal selection, when selection is based on single plants or single plots, the emphasis is on the elimination of weak and undesirable plants or clones. The breeder cannot reasonably hope to identify superior' genotypes at this stage. In the later stages, when replicated trials are the basis of selection, the emphasis is to identify and select the superior clones. The various steps involved in clonal selection are briefly described below and are depicted in figure.

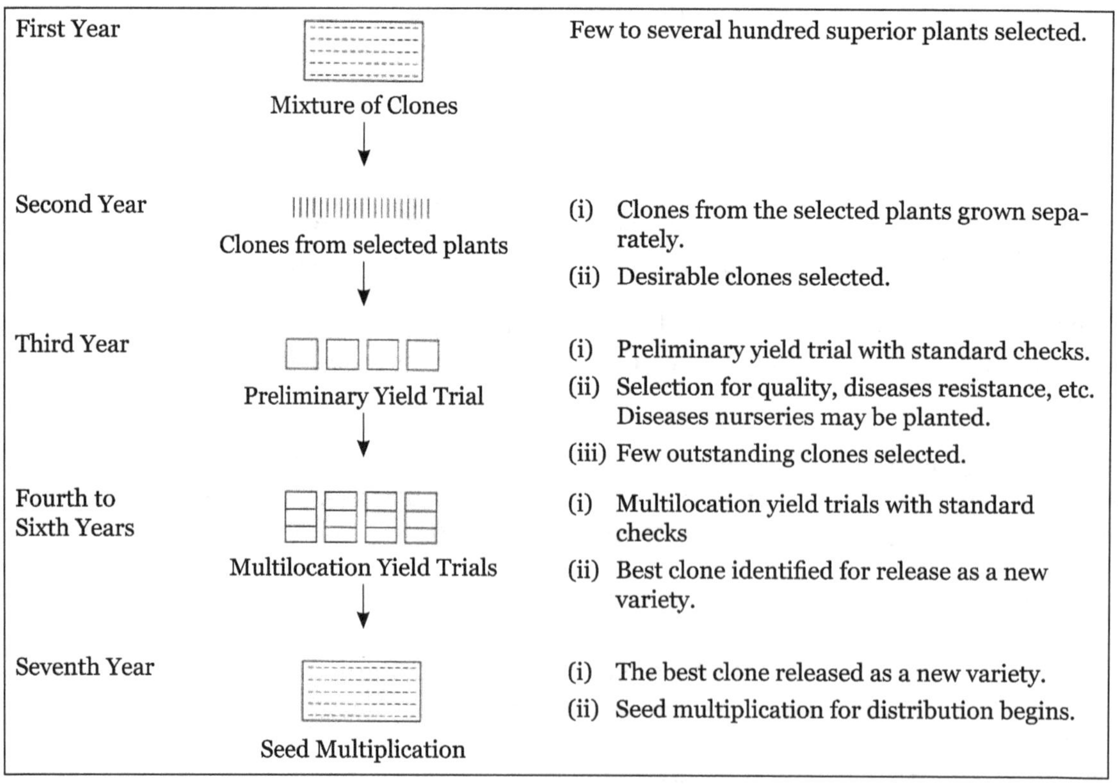

- First Year: From a mixed variable population, few hundreds to few thousand desirable plants are selected. A rigid selection can be done for simply inherited characters with high heritability. Plants with obvious weaknesses are eliminated.
- Second Year: Clones from the selected plants are grown separately, generally, without replication This is done in view of the limited supply of the propagating material for each clone, and because of the large number of clones involved. The characteristics of clones will be more clear now than in the previous generation when the observations were based on individual plants. The number of clones is drastically reduced and inferior clones eliminated. The selection is based on visual observations and on the basis of clonal characteristics. Fifty to one hundred clones may be selected on the basis of clonal characteristics.
- Third Year: Replicated preliminary yield trial is conducted. Suitable, checks included for comparison. Few superior performing clones with desirable characteristics selected for multilocation trials. At this stage, selection for quality is also done. If necessary, separate disease nurseries may be planted to evaluate the disease resistance of selected clones.
- Fourth to Sixth Years: Replicated yield trials are conducted at several locations along with a suitable check. The yielding ability, quality and disease resistance, etc. of the clones are rigidly evaluated. The best clone that is superior to the check in one or more characteristics is identified for release as a new variety.
- Seventh Year: The superior clone is multiplied released as a new variety.

Merits of Clonal Selection

- It is the only method of selection applicable to clonal crops. It avoids inbreeding depression, and preserves the gene combinations present in the clones.
- Clonal selection, without any substantial modification, can be combined with hybridization to generate the variability necessary for selection.
- The selection scheme is useful in maintaining the purity of clones.

Demerits of Clonal Selection

- This selection method utilizes the natural variability already present in the population; it has not been devised to generate variability.
- Sexual reproduction is a prerequisite for the creation of variability through hybridization.

References

- Crop-improvement-selection-methods-with-diagram, crops: biologydiscussion.com, Retrieved 19 April, 2019
- Section, naturalselection, evolution, biology: sparknotes.com, Retrieved 25 Februray, 2019
- Crop-improvement-selection-methods-with-diagram, crops: biologydiscussion.com, Retrieved 26 July, 2019

- Mass-selection-features-types-methods-crop-improvement-botany, mass-selection, breeding-methods, plant-breeding: botanylibrary.com, Retrieved 21 may, 2019
- Main-plant-breeding-methods-botany, plant-breeding: biologydiscussion.com, Retrieved 8 january, 2019
- Progeny-selection-merits-and-demerits-procedures-plant-breeding, progeny-selection, breeding-methods, plant-breeding: botanylibrary.com, Retrieved 1 Februray, 2019

4

Plant Molecular Breeding

The application of the tools of molecular biology towards plant and animal breeding is known as molecular breeding. Genetic markers such as molecular markers, morphological markers and biochemical markers are used widely in plant breeding. This chapter discusses in detail the theories and methodologies related to plant molecular breeding.

Molecular Markers

A genetic marker is a gene or DNA sequence with a known location on a chromosome that can be used to identify cells, individuals or species. It can be described as a variation which may arise either due to genetic effect (mutation or alteration in the genomic loci) or environmental effect and may represent polymorphism (genetic differences) between individuals of same or different species, in a population or between populations in a species. Generally, they do not represent the target genes themselves but act as 'signs' or 'flags'. They are located in close proximity to genes (i.e. tightly linked) may be referred to as gene 'tags'. Such markers themselves do not affect the phenotype of the trait of interest because they are located only near or 'linked' to genes controlling the trait. Like genes, all the genetic markers occupy specific genomic positions within the chromosomes and are associated with the particular trait.Such genetic maps serve several puposes, including detailed analysis of associations between economically important traits and genes or quantitative trait loci (QTLs)and facilitate the introgression of desirable genes or QTLs through marker assisted seolection. Genetic markers as well as linkage maps were described before the discovery of proteins and DNA.

Genetic markers can be divided into three main categories:

a. Morphological markers: visually accessible traits such as flower color, seed shape, growth habits or pigmentation.

b. Biochemical markers: allelic variants of enzymes called isozymes that are detected by electrophoresis and specific staining. They are able to detect diversity at functional gene level and have simple inheritance.

c. Molecular markers: based on DNA assays. Such markers themselves do not affect the phenotype of the trait of interest because they are located only near or 'linked' to genes controlling the trait.

All genetic markers occupy specific genomic positions within chromosomes (like genes) called loci. The major disadvantages of morphological and biochemical markers are that they may be limited in number and are influenced by environmental factors or the developmental stage of the plant.

However, despite these limitations, morphological and biochemical markers have been extremely useful to plant breeders. DNA markers are the most widely used type of marker predominantly due to their abundance. They arise from different classes of DNA mutations such as substitution mutations (point mutations), rearrangements (insertions or deletions) or errors in replication of tandemly repeated DNA. These markers are selectively neutral because they are usually located in non-coding regions of DNA. Unlike morphological and biochemical markers, DNA markers are practically unlimited in number and are not affected by environmental factors and/or the developmental stage of the plant

Classification of Molecular Markers

Molecular markers reveal genetic differences that can be visualized by using a technique called gel electrophoresis and staining with chemicals (ethidium bromide or silver) or detection with radioactive probes. Markers that reveal differences between any two individuals are called polymorphic markers whereas that does not discriminate between genotypes are called monomorphic markers. Polymorphic markers may also be described as co-dominant or dominant. This description is based on whether markers can discriminate between homozygotes and heterozygotes. Co-dominant markers indicate differences in size, can analyze one locus at a time whereas dominant markers indicate difference based on their presence or absence and can analyze many loci at one time. Different forms of a DNA marker (i.e. different sized bands on gels) are called marker alleles. Co-dominant markers may have many different alleles whereas a dominant marker only has two alleles.Ideal marker system should have these properties:

1. Easily available and identifiable
2. Associated with a specific locus
3. Easy and rapid assay
4. Highly variable, polymorphic and reproducible
5. Co-dominant inheritance and recurrent occurrence in genome
6. Selectively neutral to environmental conditions or management practices
7. Easy data exchange between different laboratories
8. Cost and time effective

DNA markers may be broadly divided into three classes based on the method of their detection:

1. Hybridization based
2. Polymerase chain reaction based
3. DNA sequence-based.

Essentially, DNA markers may reveal genetic differences that can be visualised by using a technique called gel electrophoresis and staining with chemicals (ethidium bromide or silver) or detection with radioactive or colorimetric probes. DNA markers are particularly useful if they reveal differences between individuals of the same or different species.

Hybridization based Markers

In hybridization based markers, DNA profiles are visualized by hybridizing the restriction endonuclease digested DNA fragment, to a labeled probe, which is a DNA fragment of known sequence e.g. Restriction fragment length polymorphism (RFLP). RFLPs are single or low copy DNA fragments and are simply inherited. These probes could be genomic clones, cDNA clones or even cloned genes. They show co-dominance and are highly reliable in linkage analysis and breeding. Their detection is based on radio-labelling, require large amounts of DNA, are labour intensive and relatively expensive and hazardous and only one marker may be polymorphic which is highly inconvenient especially for crosses between closely related species and their inability to detect point mutations and polymorphism. Hence their large scale use in practical plant breeding may be restricted. They have their DNA rearrangements due to evolutionary processes, point mutations within the restriction enzyme recognition site, mutations within the fragments and unequal crossing over.

PCR based Markers

PCR based markers involve in vitro amplification of particular DNA sequence with the help of specifically or arbitrarily chosen oligonucleotide sequences (primers) and a thermostable DNA polymerase enzyme. The amplified DNA fragments are separated by electrophoresis and banding patterns are detected by different methods such as staining (using ethidium bromide dye) and autoradiography.

- Sequence tagged sites (STS): The RFLP probes, linked to desirable traits can be converted to PCR based markers. In this, the RFLP probes are endsequenced and complementary primers are synthesized. These primers generally (20mers) are then used for amplifying specific genomic sequences using PCR.
- Expressed sequence tags (EST): These markers are developed by end sequencing of random cDNA clones. The cDNA markers are first mapped as RFLP markers and then partially sequenced to convert them into PCR based markers. Thus, these are like STS markers. These can be used for synteny mapping and cloning of specific genes.
- Simple sequence repeats (SSR): Also called the microsatellites, originally designated as short tandem repeats (STRs) dispersed throughout the genome. These are generally di- to tetra nucleotide repeats and are highly hypervariable and thus highly polymorphic. These are flanked with unique sequences that are highly conserved. The flanking unique sequences are analyzed and their complementary primers synthesized. These can thus be assayed with PCR and act as co-dominant markers. Thence, they can easily differentiate between heterozygotes and homozygotes. These are also referred to as simple sequence length polymorphism (SSLP). Allelic differences are usually a result of variable number of repeat units (VNTRs) or hypervariable regions (HVRs).
- Random amplified polymorphic DNA (RAPD): In this arbitrary decamer sequences are used as primers for amplification. These markers are dominant markers because the polymorphism is due to presence or absence of a particular amplified fragment. One major advantage of these markers is that this does not need any prior sequence informatio. One major limitation is lack of repeatability in certain cases.

- Sequence characterized amplified regions (SCAR): These markers overcome the limitations of RAPDs. In this, the RAPD fragments that are linked to a gene of interest are cloned and their termini sequenced. Based on the terminal sequences, longer primers (20-mers) are designed. These SCAR primers lead to a more specific amplification of a particular locus. These are similar to STS markers in construction and application. The presence or absence of the band indicates variation in sequences. The SCAR markers thus are dominant markers.
- Amplified fragment length polymorphism (AFLP): In this technique, restriction fragments generated by a frequent (4 base) and a rare (6 base) cutter are anchored with oligo-nucleotide adapters of new oligo-nucleotide bases. This method generates a large number of restriction fragments facilitating the detection of polymorphism. The number of DNA fragments amplified can be controlled, by choosing different base numbers and composition of nucleotides in the adapters. This technique is more reliable since stringent reaction conditions are used for primer annealing. This technique thus shows a combination of RFLP and PCR technique and is extremely useful in detection of polymorphism between closely related genotypes.
- Diversity array technology (DArT): Diversity array technology is a high throughput DNA polymorphism analysis method that combines miniaturization of microarrays and restriction based PCR. It is capable of providing comprehensive genome coverage even in organisms without any DNA sequence information. It is a solid state open platform method for analyzing DNA polymorphism. This is a high throughput relatively low cost technique. DArT has been used in whole genome sequence processing of barley and for high throughput profiling of hexaploid wheat genome.
- Cleaved amplified polymorphic sequences (CAPS): These polymorphic patterns are generated by restriction enzyme digestion of PCR products. Such products are compared for their differential migration during electrophoresis. PCR primers for this process can be synthesized based on the sequence information available in the database of genomic or cDNA sequence or cloned RAPD bands. These markers are co-dominant in nature.

DNA Sequence based Markers

Single nucleotide polymorphism (SNP): Compared to the gel based molecular marker system, SNP determination and analysis can be carried out with non-gel based assays. SNPs are DNA sequence variations that occur when a single nucleotide (A, T, C or G) in a genome sequence is changed. Most SNPs (actually 2 out of every 3 SNPs) involve the replacement of cytosine (C) with thymine (T). SNPs occurs every 100 to 300 bases along the genome. Two types of nucleotide base substitution resulting in SNPs are transition (occurs between purines-A and G) or pyrimidines (C and T) and transversion (occurs between purine and pyrimidine). Transition type of substitution constitutes 2/3rd of all SNPs and successful amplification is detected by fluorescent dyes and is allele specific. They can be generated by sequencing, single strand conformational polymorphism, aligning and comparing multiple sequences of the same region from public genome and EST databases. They can be genotyped either by allele specific hybrization, primer extension, oligo-nuclotide ligation or invasive cleavage. Its advantages include, if the exact nature and location of the allelic variation is known, large number of samples can be screened for a marker using a variety of high throughput techniques. These are likely to occur at higher frequencies in less conserved genes.

Table: Comparative study of different markers.

	Isozyme	RFLP	RAPD	AFLP	SSR	SNP
Abundance	Low	Medium	Very high	Very high	High	Very high
Types of polymorphism	Amino acid change in polypeptide	Single base change, insertion, deletion, inversion	Single base change, insertion, deletion, inversion	Single base change, insertion, deletion, inversion	Repeat length variation	Single base change
DNA quality	-	High	Medium	High	Medium	Medium
DNA sequence information	-	Not required	Not required	Not required	Required	Required
Level of polymorphism	Low	Medium	High	High	High	High
Inheritance	Co-dominance	Co-dominance	Dominance	Dominance	Co-dominance	Co-dominance
Reproducibility	Medium	High	Low	Medium	High	High
Technical complexity	Medium	High	Low	Medium	Low	Medium
Developmental cost	Medium	High	Low	Low	High	High
Automation	Low	Low	Medium	Medium	High	High

Application of Genetic Markers

Genetic markers can be used for crop improvement in several ways:

1. DNA fingerprinting for varietal identification: Inorder to protect varieties developed by originating plant breeder Plant Breeder's rights have been provided. IN order to achieve this, DNA fingerprinting can be used for varietal identification as well as for ascertaining variability in the germplasm. RAPDs, microsatellites and AFLPs are the markers of choice for the DNA fingerprinting. The fingerprinting information is useful for quantification of genetic diversity, characterization of accessions in plant germplasm collections and for protection of germplasm especially the CMS lines. These markers have been used to differentiate even closely related cultivars.

2. Germplasm evaluation: Evaluation of germplasm is important for differentiating cultivars, for construction of heterotic groups, for identification of germplasm redundancy, for monitoring genetic shifts that occur during germplasm storage, regeneration, domestication and breeding, for screening germpalsm for novel/superior genes (alleles) and for constructing a representative subset or core collection. SSRs are suitable for purity control and differentiation of plant cultivars.

3. Phylogenetic and evolutionary studies: To discern evolutionary relationships within and between species, genera or larger taxonomic groupings genetic markers are of use. Such studies involve studying similarities and differences among taxa using numerous genetic markers. Although phylogenetic trees have previously been established for many species on the basis of visible and

isozyme markers and chromosome homology, the DNA markers have recently added to length and (breadth to phylogenetic information available for a number of species. RAPDs in wheat, ISSRs in rice and RFLPs have been used more frequently for phylogenetic and evolutionary studies. RFLPs have been used for assuming the relationship between hexaploid genome of bread wheat and its ancestors.

4. Development of saturated maps: Saturated linkage maps are a pre-requisite for gene tagging, marker assisted selection and map based gene cloning. Saturated linkage maps have been developed in several crop plants like maize, rice, tomato, wheat, potato, barley, cotton, Brassica etc. In rice centromere positions in all the 12 linkage groups have been defined. The markers used for developing saturated maps include RFLPs, RAPDs, SSRs, AFLPs, SNPs, DArT and a combination of these.

5. Gene tagging: Gene tagging refers to mapping of genes of economic importance close to known markers so that gene and marker cosegregates. With the construction of molecular map, especially the RFLP maps, several genes of economic importance like disease resistance (xa 1,2,3,4,5,8,13,21,26,27,29,30 in rice), stress tolerance, insect resistance, fertility restoration genes (Rfg1 in rye), yield attributing traits etc. have been tagged.

6. Marker assisted selection: Marker assisted selection (MAS) is useful for gene pyramiding, marker assisted alien introgression and simultaneous identification and pyramiding of QTLs from primitive cultivars and alien species. RFLP markers were used for detecting introgression in interspecific backcross derived lines of the cross O. sativa (AA) x O. officinalis (CC). QTLs for yield and yield improving traits have been identified and transferred into rice O. sativa from its wild relative O. rufipogon. MAS has been utilized for incorporation of xa 13 and Xa 21 of bacterial nlight resistance in rice and two varieties have been released in India Samba Mahsuri and Pusa Basmati 1 for cultivation.

7. Comparative mapping: Molecular markers are being used extensively for studying the divergence and evolution of crop plants. In this marker clones especially cDNA clones of one crop plant are being mapped onto the linkage maps of other crops. This approach of comparative mapping is useful in several ways:

- More saturated maps can be generated by mapping marker clones of one crop onto the linkage map of other crops.
- By cross mapping, divergence and evolutionary history of various crop plants can be revealed.
- It can make gene cloning from complex organisms comparatively easier.

The first comparative maps in plants were generated in tomato, pepper and potato genomes by Tanksley et al 1988. Since then comparative maps have been generated in several crops especially in grasses, rice and wheat.

8. Map-based gene cloning: It refers to the isolation of a gene corresponding to a target trait using molecular maps. Map-based cloning consists of four major steps:

- Development of a high-resolution molecular linkage maps in the region of interest.
- Physical mapping of the region of interest by yeast artificial chromosome (YAC) or bacterial artificial chromosome (BAC) contigs.

- Identification of appropriate YAC or BAC clones for isolating putative clones harbouring the gene of interest.
- Verification through transformation that the target gene is isolated.

Plant Mutagenesis

A knowledge of the laws of mutation must sooner or later lead to the possibility of inducing mutations at will and so of originating perfectly new characters in animals and plants. And just as the process of selection has enabled us to produce improved races, greater in value and in beauty, so a control of the mutative process will, it is hoped, place in our hands the power of originating permanently improved species of animals and plants.

Importance of Variation in Plant Breeding

Variation is the source from which plant breeders are able to produce new and important cultivars. Alleles of varying forms at given loci in a population can be selected and fixed within a new individual or line. We depend on recombination and independent assortment of favorable alleles to produce new and unique individuals from which to select and produce the lines that will serve as our cultivars. With tens of thousands of genes within each crop species genome, the possibilities seem limitless. This, however, is not so clearly the case.

Consider that variation within a population can be exploited by selecting individuals with new combinations of desirable traits or alleles, most crop species, which have been selectively bred for centuries, have large portions of their genome essentially fixed. This means then, that the portions of DNA, and therefore number of traits available for re-shuffling is reduced. This reduction, however ominous it may seem, has still allowed us to make significant gains in crop yield and quality in recent decades.

Importance of Mutation in Creating Variation

So, while recombination of alleles provides offspring with presumably selectable variation for the spectrum of traits exhibited, it is only capable of creating new combinations of traits already existing. Recombination does not of itself produce novel traits. This ability is only attainable through the act of mutation, which can ultimately lead to new species. Harten gives his working definition of mutation in plants as “any heritable change in the idiotypic constitution of sporophytic or gametophytic plant tissue, not caused by normal genetic recombination or segregation”. These changes in our target plant can be passed on to progeny and used for human benefit through breeding. The occurrence of mutations within the genome of plants is rare, and in natural settings can be lethal. Through breeding and selection, beneficial mutants can be identified and used to improve target species.

Naturally Occurring Mutations

Mutations occur spontaneously in natural settings quite frequently. They can happen due to mistakes made during cell replication or exposure to mutagens such as radiation. It is estimated that a mutation occurs every 10^{-8} base pair per generation in eukaryotic genomes. In corn (Zea mays), mutations occur from 10^{-6} to 5×10^{-4} per base pair per generation. Those that we can track easily in the offspring are mutations occurring in either the gametes or cells that give rise to the gametes. Mutations in somatic cells cannot be easily tracked, nor can they be passed on to future generations, and so are only important in vegetatively propagated species.

Mechanics of Mutation

Before going any further in our discussion of mutation in plants, it is prudent to touch on the genetics behind mutations, the varying types of mutations, natural versus spontaneous, the mechanics of mutagenic agents, and the implications.

A mutation is any change within the genome of an organism that is not brought on by normal recombination and segregation. Causative agents are many, but include exposure to mutagenic agents such as radiation or certain chemicals, and mistakes made during normal cell division and replication. Most of these act upon the genome at random and are occurring all the time. These mutations are usually benign and go unnoticed in the organism due to the many cellular mechanisms that protect against these sorts of genetic mistakes. Mutations that are not caught by DNA repair mechanisms in the cell can go on to affect the organism and be present in future progeny. In Luria and Delbruck's classic experiment on bacterial resistance, they demonstrated that mutations occurred within the population at random and go unnoticed until some sort of selective pressure is placed upon the population.

Spontaneous mutations are those that occur without human intervention. These types of mutation happen randomly and the cause of them therefore not easily traceable. We do know, however, several things that cause mutations or increase their frequency. This includes the activity of mutagens found in nature such as solar radiation or reactive chemicals such as depurinators or free radicals. Mistakes during the replication of DNA during mitosis or unequal crossing-over events during meiosis can go un-repaired in a cell, leading to mutant progeny cells. It is hard to tell at what rate these occur, because the cellular machinery typically catches these mistakes. Occasionally, however, some mistakes make it permanently into the organism's genome.

Types of Mutation

At the most basic level, there are only a handful of classifications of mutations regardless of causative event. Deletions and insertions involve the removal or addition of segments of DNA respectively. These segments can range from individual base-pairs to several thousand base-pairs long. Substitutions occur when a particular base is replaced with one of the other three nucleotide bases. Inversions are instances where a segment of chromosome is rotated and replaced in the opposite direction that the segment was facing. The last major classification is a reciprocal translocation. This involves the excision of segments from two non-homologous chromosomes. These excised portions are then inserted into the other chromosome. Chromosome A will gain the segment from Chromosome B and Chromosome B gains the segment from chromosome A.

Mutagens and Implications of Each

Several types of mutagenic agents exist and have been used extensively since their discovery to produce variation and answer genetic questions. HJ Muller first discovered and used the mutagenic properties of X-ray radiation to study the genetics of Drosophila flies and the mechanics of heredity. Following this discovery, the use of ionizing radiation and chemical mutagens has been used extensively in the study of genetics and contributed a great deal to our understanding of biology as a whole.

Ionizing Radiation

Ionizing radiation includes ultra-violet (UV) light, X-ray, Gamma rays, and neutrons. These high-energy forms of radiation cause double-strand breaks of the DNA double helix. Once pieces of the DNA are broken, cellular repair mechanisms stitch the pieces back together. These DNA repair systems can only handle low rates of radiation, however, and increases in the rate of exposure to ionizing radiation causes permanent mutations to occur and accumulate in an organism's genome. Radiation causes deletions of nucleotides from the DNA sequence. These deletions can cause reading-frame shifts, inactive protein products, or faulty transcripts. This typically results in null mutations, which are those in which a particular gene is inactivated.

Chemical Mutagens

Chemical mutagens affect the DNA molecule through chemical reactions within the genome. Base analogs are chemicals with similar properties to the DNA bases. They can be incorporated by the cell into the genome, replacing the proper base. Alkylating agents such as ethyl methanesulfonate (EMS), react with guanine or thymine by adding an ethyl group which causes the DNA replication machinery to recognize the modified base as an adenine or cytosine, respectively. This base substitution typically does not result in reading frame shifts, but instead causes altered forms of a triplet sequence. Changing a single base within a coding region causes either a nonsense codon which stops transcription or an altered codon which changes the amino acid transcribed, which can de-activate, reduce efficiency of, or produce a new protein. Nitrous acid, a deaminating agent, removes the amine group from adenine or cytosine. When the cell replicates this altered area, it matches adenine to the deaminated cytosine, and cytosine to the deaminated adenine, resulting in similar effects to that of alkylating agents. The last type of chemical mutagen, intercalating agents, causes deletions, reading frame shifts, or random base insertions. These compounds insert themselves into the DNA between adjacent base pairs, thus disrupting replication and transcription machinery.

Transposable Elements

Transposable elements are a special class of mutagen. They are self-replicating segments of DNA that excise and/or insert themselves within the genome. Also known as transposons, these strange sequences were first proposed by the pioneering Barbara McCLintock working on maize. Transposable elements, unlike other forms of mutagenesis, do not act upon the genome in a completely random fashion. Rather, they have certain "hot-spots" where they are more likely to insert or replicate themselves. By their insertion or deletion, they act upon the genes in which they are located or those adjacent to them. Transposable elements can cause gene disruptions, protein product alterations, or large-scale genome rearrangements. If inserted into the intron of a gene, they can cause transcriptional inefficiency.

Use of Mutation in Plant Breeding

Mutagenesis, the act of inducing mutations within an organism's genome, has been used in plant breeding since Muller's discovery of the mutagenic effects of X-rays on Drosophila flies. The first crop species to be mutagenized was barley by LJ Stadler, who began using X-rays to induce mutations independently of Muller at around the same time, although he published his first paper on the subject the following year. These early mutation experiments were designed mainly to discern genetic truths of inheritance and chromosomal theory. Recall that at this point Watson and Crick's double-helix DNA model had still not been arrived at.

Dose, Rate, Species/Genotypes, Conditions of application, etc. Much of the early work done with ionizing radiation and chemical mutagens was an effort to determine efficient doses and exposures of the various agents to effect high percentages of mutations without causing lethality. The researchers noticed that the rates and doses varied tremendously for species, genotype, ploidy level, and the conditions in which treatment were conducted. Grays (Gy) are the measurement unit of radiation dose. For the sterilization of food products, processors typically use rates as high as 10 kGy. In the mutagenic treatment of plant material, doses can range from as low as 2 Gy for cell cultures or leaf tissues, to as high as 700 Gy for seed material.

When determining the most efficient dose for one's crop species, it is important to first consult the literature for any information on the mutagenic agent to be used and the crop it will be used on. Chances are good that someone has already used the particular mutagenic agent on your crop. If not, it is important to consider several factors:

1. The first of course is safety issues regarding the mutagen. A researcher needs to know what sort of certifications, licenses, and precautions must be taken while carrying out the experiment and handling the plant material after the experiment is complete.

2. Next is the mutagen's type and mechanism. Knowing how the mutagen affects the experimental tissue will not only allow for greater safety, but will also help in making better decisions on how to carry-out the experiment. For instance, if using X-rays, distance from the source of radiation reduces dose. It would be unwise to place a bucket of seed in front of an X-ray source and assume all the seeds in the bucket received equivalent doses of radiation. Similarly, the mutagenic chemical, EMS, requires very stringent experimental conditions to effect mutations.

3. Tissue specific reactions. Seeds react differently than stem cuttings, meristem tissues or callous tissue to the same level of radiation.

4. Species and ploidy level affect mutational response to the mutagen a great deal. Some crop species are equipped to handle a greater mutational load than others. They can more efficiently repair damage to their DNA. Ploidy level of the target species also influences mutation response.

Screens and Efficiency

Recall that most agents used to induce a mutagenic response act upon the target genome in a random fashion. There is no precise control of where mutations will occur within the genome, within particular cells, or to what degree or number of mutations will occur within the target genome. To limit or maximize the number of mutations occurring, it is possible to adjust the dosage of mutagen treatment to the desired effect. For instance, if the mutation response desired is a high level, the researcher will expose the experimental material to high doses, which will result in high lethality of plants, but a high mutational response. Lower doses will result in fewer mutations per genome, a higher survival rate, and possibly also a higher level of non-mutant recovery. The purposely ambiguous terms low and high are used here due to the variable nature of mutation experiments and the large number of factors involved. For example, Auld exposed germinating cotton seeds to 3 to 5 times the LD50 rate of EMS (3% v/v) to ensure recovery of mutants in his tetraploid Gossypium hirsutum subjects. In contrast, EMS treatment of germinating sugar beet seeds only required 0.5% v/v EMS to elicit a sufficiently mutagenic response.

Once the desired dose-response relationship has been established in the experimental materials, the next step is the implementation of the "screen." The first level of the screen is simple, the M1 generation (M notation denotes generations after mutagenic treatment, so the parent is the M0 and the first generation after mutagenesis is the M1 generation) must be grown out. Most mutant alleles are recessive, so successful mutants would not be seen until the M2, at which point lethality and infertility will often significantly affect the plants in the population. So the first level of the screen is identifying viable mutants.

Next, a cheap, fast, high throughput method of phenotyping and screening for the trait of interest must be in place. This is the most important factor in any mutagenesis experiment. Due to the chimeric nature of mutation, it is likely that some of the viable plants seen in the M2 population will be non-mutagenized parent types. Identification of the desired mutants requires looking at and accurately identifying the mutants. Accurate measurements must be fast and affordable to process the necessary number of individual plants or lines to achieve success. Like all plant breeding methods, mutation breeding is also a numbers game.

In the sugar beet experiment noted above, of over 3200 M2 families derived from an early bolting line treated with EMS, only 9 families exhibited the desired non-bolting trait and eventually gave rise to 5 lines with the non-bolting phenotype. The success of this program was in its simple and efficient screen. The experimenter observed the M2 population for non-bolting visually. Many other traits are not so simply characterized, but can still be done fairly cheaply and quickly.

Let us use the cotton example from above. Auld treated 2 kg of germinating cotton seeds with EMS and the fiber from nearly 2000 individual M3 plants characterized using the High Volume Instrument (HVI). This machine measures cotton fiber quality but is considered less accurate than other methods. HVI measurement, however, is cheap, fast, and highly correlated to yarn and textile performance. Although not perfectly accurate, it is good enough for processing large numbers. These 2000 M3 individuals gave rise to 2 M4:5 lines with fiber lengths that exceeded the parent value by 10%. One out of every thousand plants was a desirable mutant.

Mutation Breeding in Self Fertilizing Species

Breeding mutant traits is fairly straightforward in crops that are capable of self-fertilization. Because many mutations are recessive, after mutagenic treatment, the material should be self-fertilized and advanced to at least the M2 before phenotypic screening. At this point plants will be segregating for the recessive mutant trait. Positive mutant identifications should be kept for future selection. Because mutagens act randomly upon the genome, it is important to collect as many positive mutants as possible. This allows the breeder to have a series of lines from which to select for performance in addition to the presence of the mutant trait.

Mutation Breeding in Cross Fertilizing Species

Cross-fertilizing species raise some difficulties. Because species which are predominantly cross-fertilizing typically exhibit significant inbreeding depression, the necessary self-fertilizations to identify mutants in the population result in reduced plant vigor due to the genetic background and not necessarily the mutations. This compounds the difficulty of successfully identifying mutations. Dominant mutations can be identified, but these occur very rarely. Crop species with self-infertility mechanisms are especially hard to use mutation breeding methods without elaborate crossing schemes. The numbers required to make this feasible, however, make this impractical if not essentially impossible.

Mutation Breeding in Vegetatively Propagated Species

When attempting to effect mutation in vegetatively propagated species such as sugarcane or banana, it is important to note the chimeric nature of mutagenic treatment. All cells exposed to the mutagen will not necessarily incur mutations, but those that do incur mutations, will give rise to cells exhibiting the mutation. For this reason it is important to treat parts of the plant that will give rise to either seed or vegetative propagules. Identification and propagation of the necessarily large numbers of plants to identify successful mutants is difficult for many vegetatively propagated plants, however, once one is identified, the mutation is fixed in the cloned progeny. Crop species where in vitro techniques exist and can be used to mutate plant material, allows for the regeneration of large numbers of plantlets. This system is highly amenable to both vegetatively and seed propagated species.

Mutation Breeding in Seed Propagated Species

Seeds treated with mutagenic agents give rise to chimeric plants. Chimeric plants produce both mutant and non-mutant seed. This can be problematic; however, one just needs to plant more seeds to find the desired mutants. As long as an efficient screening method is in place, this should produce no significant pitfalls. Mutagenic treatment of seed is by far the most popular method in mutation breeding programs.

Ploidy and How it affects Mutation Breeding

Mutagenesis of polyploid plant species is difficult. Because most mutations are recessive, plants must be homozygous to display the trait. Polyploid conditions can further complicate the process of reaching homozygosity for the mutation, so must be selfed for additional generations to ensure presence of the mutation.

Polyploidy in Plant Breeding

Polyploids are organisms with multiple sets of chromosomes in excess of the diploid number. Polyploidy is common in nature and provides a major mechanism for adaptation and speciation. Approximately 50-70% of angiosperms, which include many crop plants, have undergone polyploidy during their evolutionary process. Flowering plants form polyploids at a significantly high frequency of 1 in every 100,000 plants. Many studies have been carried out to understand the nature of polyploidism. To understand polyploidy, a few basic notations need be defined. The basic complete set of chromosomes is designated by "x" while the total number of chromosomes in a somatic cell is designated "2n". The total number of chromosomes in a somatic cell is twice the haploid number (n) in the gametes. The ploidy of some of the major crops in the world is represented in table.

Classification of Polyploids

Polyploids may be classified based on their chromosomal composition into either euploids or aneuploids. Euploids constitute the majority of polyploids.

Euploidy

Euploids are polyploids with multiples of the complete set of chromosomes specific to a species. Depending on the composition of the genome, euploids can be further classified into either autopolyploids or allopolyploids. Tetraploidy is the most common class of euploids.

Table: A list of major crops and their ploidy.

Common name	Ploidy	Name	Propagation Maize
Maize	2x=20	Diploid	Outcrossing
Wheat	6x=42	Hexaploid	Outcrossing
Rice	2x=24	Diploid	Selling
Potatoes	4x=48	Tetraploid	Outcrossing Vegetative
Soybeans	2x=40	Diploid	Selfing
Barley	2x=14	Diploid	Selfing
Tomatoes	2x=24	Diploid	Selling
Bananas	3x=33	Triploid	Vegetative
Watermelon	2x=22	Diploid	Outcrossing
Sugarcane	8x=80	Octoploid	Outcrossing vegetative
Sugar beet	2x=18	Diploid	Outcrossing
Cassava	2x=36	Diploid	Outcrossing;Vegetative

Autopolyploidy

Autopolyploids are also referred to as autoploids. They contain multiple copies of the basic set (x) of chromosomes of the same genome Autoploids occur in nature through union of unreduced gametes and at times can be artificially induced.

Natural autoploids include tetraploid crops such as alfafa, peanut, potato and coffee and triploid bananas. They occur spontaneously through the process of chromosome doubling. Chromosome doubling in autoploids has varying effect based on the species. Spontaneous chromosome doubling in ornamentals and forage grasses has led to increased vigour. For instance, ornamentals such as tulip and hyacinth, and forage grasses such as ryegrasses have yielded superior varieties following spontaneous chromosome doubling. Due to the observed advantages in nature, breeders have harnessed the process of chromosome doubling in vitro through induced polyploidy to produce superior crops. For example, induced autotetraploids in the watermelon crop are used for the production of seedless triploid hybrids fruits. Such polyploids are induced through the treatment of diploids with mitotic inhibitors such as dinitroaniles and colchicine. To determine the ploidy status of induced polyploids, several approaches may be used. These include, chloroplast count in guard cells, morphological features such as leaf, flower or pollen size (gigas effect) and flow cytometry.

Allopolyploidy

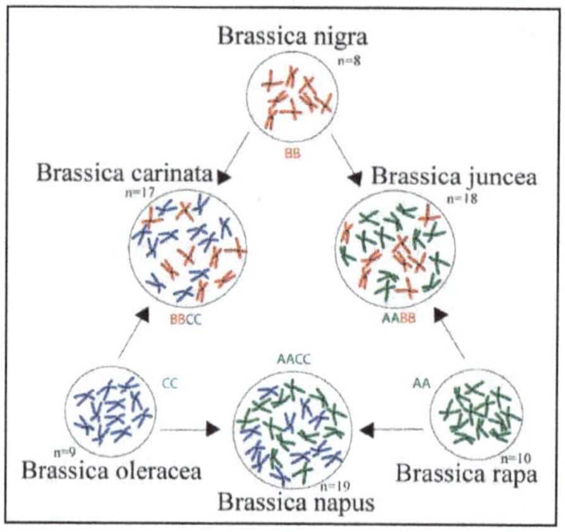

Triangle of U showing the origin of cultivated mustard.

Allopolyploids are also called alloploids. They are a combination of genomes from different species. They result from hybridization of two or more genomes followed by chromosome doubling or by the fusion of unreduced gametes between species. This process is key in the process of speciation for angiosperms and ferns and occurs often in nature. Economically important natural alloploid crops include strawberry, wheat, oat, upland cotton, oilseed rape, blueberry and mustard. To differentiate between the sources of the genomes in an alloploid, each genome is designated by a different letter. For example, the origin of the cultivated mustards (Brassica spp) has been well explained by Nagaharu in the triangle of U with each species represented by a distinct letter.

The hybridized genomes differ in their degree of homology with some being able to pair during mitosis and/or meiosis while others not. When only segments of the chromosomes of the combining genomes differ, the phenomenon is called segmental alloploidy. These chromosomes are similar but not homologous and are called homeologous chromosomes. Such chromosomes indicate ancestral homology. Induced alloploidy is not common. However, it has been used in some Genus such as Cucumis to elucidate the molecular mechanisms involved in diploidization (tendency of polyploids to act as diploids). In this study, an allotetraploid was induced by hybridization between Cucumis sativus and Cucumis hystrix followed by chromosome doubling. Cytogenetic studies were carried out in the following generations to establish the molecular mechanisms involved stabilization of newly formed allopolyploids which include neo-functionalization and sub-functionalization.

Aneuploidy

Aneuploids are polyploids that contain either an addition or subtraction of one or more specific chromosome(s) to the total number of chromosomes that usually make up the ploidy of a species

Aneuploids result from the formation of univalents and multivalents during meiosis of euploids. For example, several studies have found that 30-40% of progeny derived from autotetraploid maize are aneuploids. With no mechanism of dividing univalents equally among daughter cells during anaphase I, some cells inherit more genetic material than others. Similarly, multivalents such as homologous chromosomes may fail to separate during meiosis leading to unequal migration of chromosomes to opposite poles. This mechanism is called non-disjunction. These meiotic aberrances result in plants with reduced vigor. Aneuploids are classified according to the number of chromosomes gained or lost as shown in table.

Table: Classification of aneuploids.

Term	Chromosome number
Monosomy	2n-1
Nullisomy	2n-2
Trisomy	2n+2
Tetrasomy	2n+2
Pentasomy	2n+3

Mechanisms of Polyploidy Formation

Several cytological mechanisms are known to spontaneously induce polyploidy in plants. One such route involves non-reduction of gametes during meiosis a process called meiotic nuclear restitution. The formed gametes (2n) contain the somatic nuclear condition of cells. Meiotic aberrations related to spindle formation, spindle function and cytokinesis have been implicated in this process. The subsequent union of reduced and non-reduced gametes leads to the formation of polyploids. For example, autotetraploids may be formed in a diploid population through the union of two unreduced 2n gametes as was found in the F1 progenies of open-pollinated diploid apples. Similarly, spontaneous allotetraploids were formed in 90% of F2 progenies of interspecific crosses between Digitalis ambigua and Digitalis purpurea, which are common ornamental plants. Another example is the formation of autohexaploid Beta vulgaris (sugar beet) and alfalfa from cultivated autotetraploid varieties apparently from the union of reduced (2x) and unreduced (4x) gametes.

Another major route for polyploid formation is through somatic doubling of chromosomes during mitosis. In nature, the formation of polyploids as a result of mitotic aberrations has been reported in the meristematic tissue of several plant species including tomato and in non-meristematic tissues of plants such as bean. Artificial inducement of polyploids through the inhibition of mitosis is routine in plant breeding. High temperatures above 40°C have been used to induce tetraploid and octoploid corn seedlings albeit with low success of 1.8% and 0.8% respectively. Currently, chemical mitotic inhibitory agents such as colchicine or dinitroanilines are used to induce polyploidy in crop plants. A typical example is the production of tetraploid watermelon plants for the production of seedless triploid watermelon.

In addition, an uncommon mechanism of polyploid formation involves polyspermy where one egg is fertilized by several male nucleuses as commonly observed in orchids The major pathways involved in polyploidy formation are represented in figure.

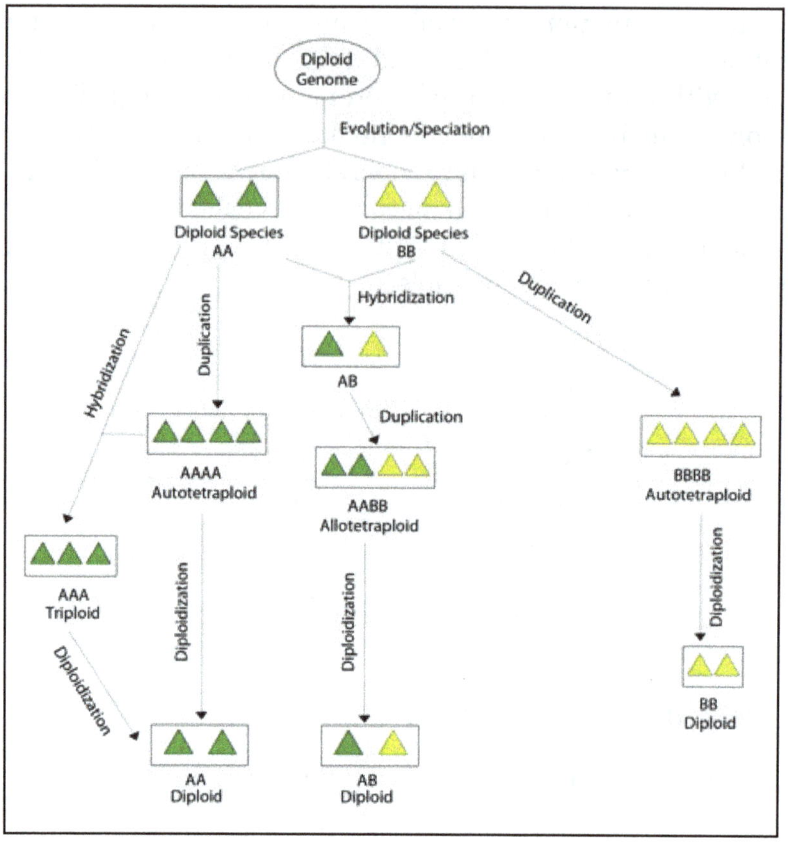

Major pathways in the formation of polyploids

Alterations Associated with Polyploidy

Several changes in the plant accompany spontaneous or induced polyploidy. These may be changes in genetic composition, physiological mechanisms, structural composition and vigor. Some of these changes create the platform for the commercial exploitation of polyploids. Genetic changes following genome duplication involve the rapid loss of chromosomal segments in a process called diploidization. Diploidization describes the process by which a polyploid genome become more 'diploid-like' in character. It is necessary to eliminate duplicated genes in a newly formed polyploid to avoid gene silencing as well as to stabilize fertility. Duplicated genes that are retained often undergo subfunctionalization (complementing genes) and neofunctionalization (genes with novel functions). Diploidization has been described for many genus including Nicotina and Cucumis.

The increase in nuclear ploidy affects the structural and anatomical characteristics of the plant. In general, polyploidy results in increased leaf and flower size, stomatal density, cell size and chloroplast count. These phenomena are collectively referred to as the gigas effect. Its application in forage and ornamental breeding is described later.

Physiological changes are also known to accompany genome duplication. These mainly result from change of metabolism resulting in a general increase in secondary metabolites.This property has found application in the breeding of medicinal herbs in the production of pharmaceuticals. Hybrid vigor resulting from interspecific crosses in allopolyploids is one of the most exploited advantages of polyploid in plant breeding.

A comparison between the leaf and flower of a (A) diploid and (B) induced tetraploid watermelon illustrating the gigas effect.

Implications of Ploidy in Plant Breeding

Heterosis in Allopolyploids

Heterosis or hybrid vigor is the difference between the hybrid and the mean of the two parents and is characterized by increased vigor and superior qualitative or quantitative traits. Over the last several decades, breeders have increased the world food production by utilizing the concept of heterosis in hybrid cultivars. For example, following the introduction of hybrid corn (diploid) in the 1920's, there was a six fold increase in corn production between then and 1990 in the U.S. However, unlike diploids which may lose heterosis with each consecutive generation due to segregation, alloploidy and autoploidy imposes pairing of homologous chromosomes, thus preventing intergenomic recombination. This concept is called preferential or selective pairing and is the tendency for a doubled set of chromosomes to pair independently of the doubled set of chromosomes of the other species. In this way, heterozygosity is maintained throughout generations. Generally, the parents used in hybrid formation should be within subspecies or between subspecies. An example of a man-made interspecies allopolyploid hybrid is triticale. It is derived from crossing two cereals, hexaploid bread wheat (T. aestivum) and rye (Secale cereale). Triticale was developed to combine good qualities of wheat including high yield and grain quality with the hardiness (disease and stress tolerance) of rye.

The process of hybrid formation for polyploids is not without setbacks. Many interspecific hybrids have low fertility and viability due to hybrid incompatibilities. Hybrid incompatibility results from genes that are functionally diverged in the respective hybrid forming species. This may lead to silencing of protein encoding genes and has been reported in interspecific hybrids of Arabidopsis. To increase the heterosis, fertility and viability of interspecific hybrids, several factors should be considered. The parents used should be of diverse genetic background and preferably heterozygous.

Inbreeding in Polyploids

Self pollination is an important method for attaining homozygosity in breeding. Through this process, it is possible to fix desired alleles in the background of a crop. In general, it takes approximately

3.80 more generations for an autotetraploid to reach the same level of homozygosity as the diploid. Fixing a trait controlled by a single gene in an autotetraploid, would require four identical alleles to achieve homozygosity. For example, in a segregating, tetraploid F_2 population the proportion of the homozygous loci would be 1:18.

Effect of Polyploidy on inheritance and Population Genetics

An immediate consequence of polyploidy is the change in gametic and filial frequencies. This is because polyploids have multiple alleles associated with a single locus. For example, a hexaploid has six alleles per locus while a tetraploid has four. The genetics of polyploids is often complicated by multi-allelism at loci thus altering segregation ratios and inheritance patterns expected in diploids. Provided a polyploid species behaves like a diploid at meiosis through normal bivalent pairing (disomic inheritance), such as in wheat or tobacco, normal biometric analysis of inheritance apply. However, several autotetraploid crop plants including potatoes, coffee and lucerne and some forage grasses have tetrasomic inheritance. With this knowledge, it is necessary to make accommodations in population structure and breeding strategy to account for differences in gamete structure. For example, breeding schemes that maximize heterozygosity are frequently used for the autotetraploid alfalfa in an attempt to utilize multi-allelic interactions. Altered genotypic ratios are apparent in polyploids when compared with diploids. For example an arbitrary locus with B (dominant) and b (recessive) alleles, following selfing, an autotetraploid (BBbb) would produce 5 possible genotypes while a diploid (Bb) would generate 3 possible genotypes. Distinguishing a quadruplex (BBBB) from a triplex (BBBb) in the segregating population using a progeny test presents difficulty in breeding because both would breed true to the dominant allele. An extra generation would be required to identify the triplex by observing the formation of duplex plants.

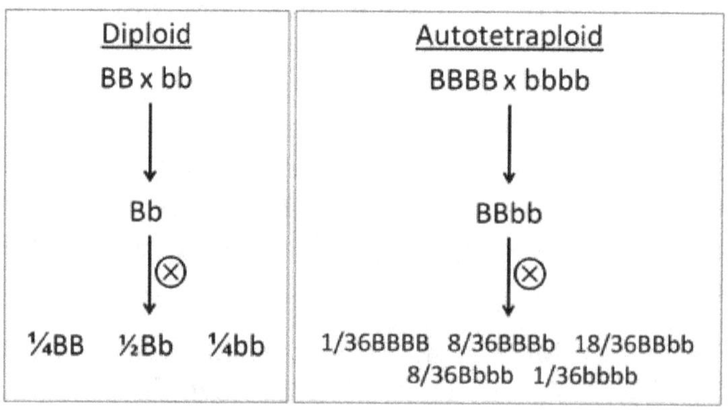

Genotypes and ratios formed following selfing of a heterozygous diploid and autotetraploid.

Effect of Polyploidy on Sterility

Since autoploids contain more than two homologous chromosomes, meiosis results in the formation of univalents and multivalent, unlike in diploids where bivalents are usually formed. For instance during meiosis, autotetraploids may form bivalents, quadrivalents and univalents. The ratio of these gametes following meiosis determines the fertility of a polyploid individual. Univalents and trivalents result in non-functional sterile gametes and are the most common in triploids, making them sterile.

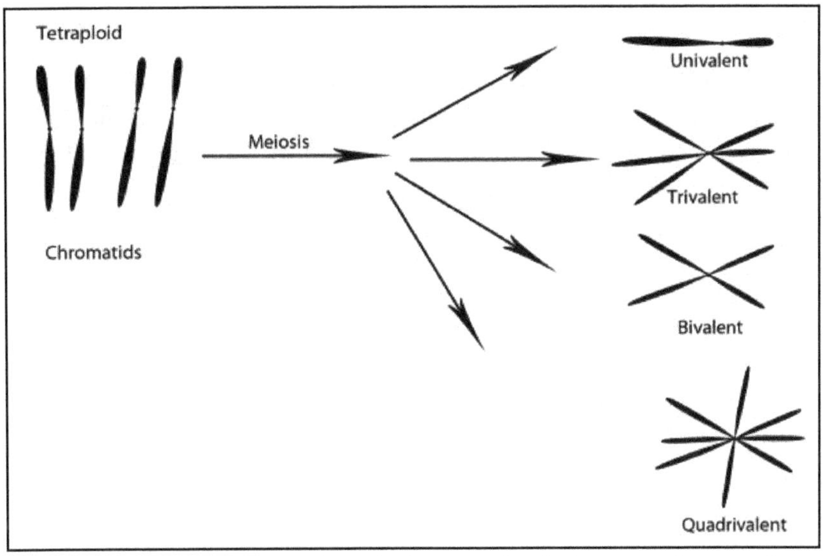

Gamete formation in autotetraploids. Trivalents and univalents result in sterile gametes while bivalents and quadrivalents result in fertile gametes.

Rigorous and effective selection strategies for fertile autoploids are practiced in the development of inbred lines. Breeders rogue out autoploids with low seed set as well as those with morphological abnormalities. Sterile alloploids arise from the pairing of homeologous chromosomes from separate genomes during meiosis instead of homologous chromosome. This results in non-functional gametes. A viable allopolyploid requires a diploid-like meiosis behavior to establish disomic inheritance and full fertility. Fertility problems in allopolyploids also occur when crossing crops of different ploidy levels as a result of formation of multivalents. To improve fertility, breeders use the parent with the lowest chromosome number as the female parent so as to increase seed set.

Common Applications of Ploidy in Crop Plants

Mutation Breeding

High frequencies of chromosome mutations are desirable in modern breeding techniques, such as tilling, as they provide new sources of variation. The multiallelic nature of loci in polyploids has many advantages that are useful in breeding. The masking of deleterious alleles, that may arise from induced mutation, by their dominant forms cushions polyploids from lethal conditions often associated with inbred diploid crops. This concept has been instrumental in the evolution of polyploids during bottlenecks where there is enforced inbreeding. Mutation breeding exploits the concept of gene redundancy and mutation tolerance in polyploid crop improvement in two ways. First, polyploids are able to tolerate deleterious allele modifications post-mutation, and secondly, they have increased mutation frequency because of their large genomes resulting from duplicated condition of their genes. The high mutation frequencies observed with polyploids may be exploited when trying to induce mutations in diploid cultivars that do not produce enough genetic variation after a mutagenic treatment. This approach has been used in mutation breeding of Achimenes sp. (nut orchids) by first forming autotetraploids through colchicine treatment followed by the application of fast neutrons and X-rays. In this study, the autotetraploids were found to have 20-40 times higher mutation frequency than the corresponding diploid cultivar due to the large genome.

Seedless Fruits

The seedless trait of triploids has been desirable especially in fruits. Commercial use of triploid fruits can be found in crops such as watermelons and are produced artificially by first developing tetraploids which are then crossed with diploid watermelon. In order to set fruit, the triploid watermelon is crossed with a desirable diploid pollen donor.

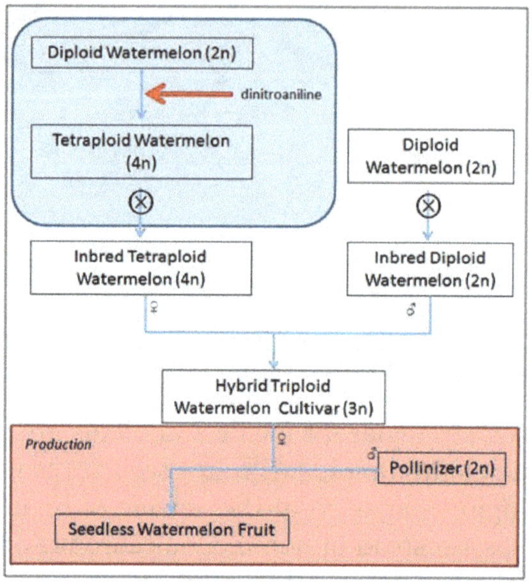

A flow diagram showing the production of seedless triploid watermelon.

Bridge Crossing

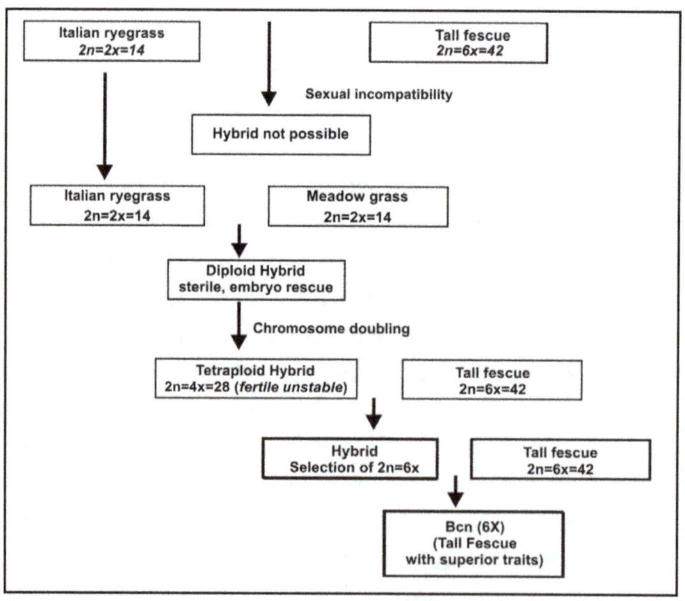

The development of superior tall fescue grass through bridge crossing and induced tetraploidy.

Another breeding strategy that utilizes the reproductive superiority of polyploids is bridge crossing. When sexual incompatibilities between two species are due to ploidy levels, transitional crosses can be carried out followed by chromosome doubling to produce fertile bridge hybrids. This method has been used to breed for superior tall fescue grass (F. arundinacea) from Italian ryegrass (2n=2x=14) and tall fescue (2n=6x=42) by using meadow grass (Fescue pratensis) as a bridge species. The same principle has been applied in fixing heterozygosity in hybrids by doubling the chromosomes in the superior progeny.

Ornamental and forage Breeding

One of the immediate and obvious consequences of polyploidy in plants is an increase in cell size which in turn leads to enlarged plant organs, a phenomenon termed gigas effect. For example, the volume of tetraploid cells usually is about twice that of their diploid progenitors. The increase in cell volume however is mainly attributed to increased water and not biomass. Therefore, its application is limited for breeding agronomically important crops such as cereals. Although chromosome doubling may result in significantly larger seeds and increased seed-protein content in cereal crops, this advantage is offset by low seed set. In contrast, the gigas effect has been explored in tree, ornamental, forage crop and fruit breeding. For example, through induced polyploidy, breeders have developed Bouschet tetraploid grapes with more yield and juice content than the diploid progenitor Alicante. Ornamental crops such as snapdragons and marigolds have been bred through chromosome doubling to improve the quality and size of their blossoms. A strong inverse correlation between DNA content and development rates in plants has been reported by several authors. It has been attributed to lower auxin levels, reduced surface to volume ratio and altered nuclear surface to cell volume ratio. The slower growth rate of polyploids allows them to flower later and for a longer period of time than their diploid progenitors. This quality may be of interest especially in ornamental breeding.

Production of Apomictic Crops

Apomixis provides another avenue for use of polyploids in breeding. Apomixis provides an avenue for the production of seeds asexually through parthenogenesis. Most apomictic plants are polyploid but most polyploid plants are not apomictic. In plants capable of both sexual and asexual reproduction, polyploidy promotes the latter. Obligate apomicts are the most desired of hybrids but little gain has been realized towards their development. However, it has been suggested that obligate apomicts may be induced through development of very high ploidy plants. An example of an obligate apomict achieved at high ploidy level is the octoploid of the grass, Themeda triandra.

Disease Resistance through Aneuploidy

Aneuploidy has been applied in breeding to develop disease resistant plants through the addition of an extra chromosome into the progeny genome. An example is the transfer of leaf rust resistance to Tricum aestivum from Aegilops umbellulata through backcrossing. In addition, other breeding strategies utilizing aneuploidy have been explored including chromosome deletion, chromosome substitution and supernumerary chromosomes.

Industrial Applications of Polyploidy

Chromosome doubling is reported to have an apparent effect on many physiological properties of a plant. The most discernable of these has been the increase in secondary as well as primary metabolism. The resulting increase in secondary metabolites, in some cases by 100%, after chromosome doubling has been widely exploited in the breeding of narcotic plants such as Cannabis, Datura and Atropa. In vitro secondary metabolite production systems that exploit polyploidism have also been developed. The production of the antimalarial sesquiterpene artemisinin has been enhanced six fold by inducing tetraploids of the wild diploid Artemisia annua L. (clone YUT16). In addition, commercial synthesis of sex hormones and corticosteroids has been improved significantly by artificial induction of tetraploids from diploid Dioscorea zingiberensis, native to China. Attempts have been made to improve the production of pyrethrin, a botanical insecticide, by chromosome doubling of Chrysanthemum cinerariifolium. Other plants whose production of terpenes has increased following artificial chromosome doubling include Carum cari, Ocimum kilmandscharicum and Mentha arvensis. The enhanced production of secondary metabolites such as alkaloids and terpenes in polyploids may concurrently offer resistance to pests and pathogens. Experiments with diploid Glycine tabacina, a forage legume, and its tetraploid forms to measure resistance to leaf rust, Phakopsora pachyrhizi, established that 42% of the tetraploid plants were resistant compared to 14% of the diploid plants. Similar results were observed while comparing resistance to insects and the clover eel nematode between Trifolium pratense (red clover) tetraploids and diploids.

5

Plant Breeding: Methods and Resistance for Abiotic Stresses

The negative impact of non-living elements on a living organism in a particular environment is known as abiotic stress. Plants are exposed to several kinds of abiotic stress such as extreme temperatures, droughts, floods and high salinity. The chapter closely examines the key concepts of plant breeding to provide an extensive understanding of the different methods which are used to increase resistance for abiotic stress in plants.

Breeding Methods for Self-Pollinated Crops

The breeding methods used for self-pollinated crops are Individual plant progeny is selected from F_2 and subsequent generations, and their progenies are tested. During this process the record of parents as well as off-springs is kept, for which it is known as pedigree method.

The pedigree is defined as the description of the ancestors of an individual and it is generally helpful in finding out the amount of relatedness among two individuals, i.e., whether they are related by common parent in their descent ancestry or not.

Procedure

First Year

The hybridization is done among two selected parents, after emasculation one become female parent and another male parent. After seed set and maturation, the F_1 seeds are harvested separately from each plant individually. On the basis of choice of parents, the type of cross will be of two types – it will be simple cross or complex cross.

Second Year

F_1 generation seeds are space planted and selfing is allowed, each F_1 will produce more F_2 seeds. From 15-30 selected F_1 plants, the F_2 seeds are collected to get a reasonable size of F_2 population and variation.

Third Year

In F_2 generation, 2000-10000 plants are space planted, 100-500 plants are selected and their seeds are harvested separately. If the parent plants are closely related varieties then the number of selected F_3 plants would be smaller whereas in case of distantly related varieties the number of F_3 progenies will be of relatively larger numbers.

Fourth Year

In F_3 generation also, the individual plant progenies are space planted. Each progeny should have about 30 or more plants. Individual plants with desirable characteristics are selected; disease and lodging susceptible progenies to be eliminated, and also the progenies with undesirable characters are rejected even from the selected plants. During this selection if the number of superior progenies is very small then the whole cross programme may be rejected.

Fifth Year

The selection procedure is same as previous year, only if two or more progenies coming from the same F_3 progeny are similar and comparable, then only one may be saved and others may be rejected. The emphasis is given on the selection of desirable plants from superior progenies.

Sixth Year

Individual plant progenies of F_5 generation are planted according to recommended commercial seed rate. Three or more rows for each progeny will help in comparison among progenies. Many progenies may have become reasonably homozygous genotype and may be harvested in bulk. If the progenies show variation then the individual plants are selected. The number of selected progenies should be reasonable so that preliminary yield trial with 25-100 progenies can be done.

F_1 generation seeds are space planted and selfing is allowed, each F_1 will produce more F_2 seeds. From 15-30 selected F_1 plants, the F_2 seeds are collected to get a reasonable size of F_2 population and variation.

Seventh Year

Individual plant progenies of F_6 generation are planted in multi-row plots and evaluated visually. Progenies harvested in bulk since they become homozygous. The segregating progenies may be discarded and the preliminary yield trial may be done for the progenies which are reasonably homozygous and have enough seeds.

Eighth Year

Preliminary yield trial with three or more replications is conducted to identify few superior lines. The progenies are evaluated for plant height, lodging, disease resistance, flowering time, maturity time, etc. Quality test is done to serve as an additional basis for selection.

Ninth to Tenth or Thirteenth Year

The superior lines are tested in replicated yield trials at several locations. The above mentioned criteria are evaluated for these lines. The line which is superior to the best commercial variety may be released as new variety.

Eleventh or Fourteenth Year

The selected strain should get multiplied to release as a new variety. Breeder has the responsibility to supply the seeds to the state seeds corporation for production and marketing of the seeds.

Year	Generation	Procedure
First Year	P_1 P_2	Selected plants of parental variety are hybridized to get the F_1 hybrid seeds.
Second Year	F_1	The hybrid seeds (10-30) are planted and the seeds of F_2 harvested a bulk.
Third Year	F_2	2000-1000 seeds are space planted. 100-500 superior plants are selected and seeds are harvested from those as pedigree.
Fourth Year	F_3	Individual plant progenies are planted in row and space planted. Superior plants are selected.
Fifth Year		The same procedure as in fourth year.
Sixth Year	F_5	The individual selected plant progenies are in multi-row plots. Superior plants are selected from the progenies.
Seventh Year	F_6	The same procedure as in previous year. Enough seeds to be collected from superior plants.
Eighth Year	F_7	The preliminary yield trial and quality test has to be done.
Ninth to Thirteenth Year	F_8-F_{12}	Co-ordinated yield trials to be done. Disease and quality tests have to be done.
Fourteenth Year	F_{13}	Multiplication of seeds for distribution and release as a new variety.

Merits

1. This method is most useful as transgressive segregation for yield and other quantitative characters may be recovered in addition to improvement of specific characters.
2. This method is well suited for improvement of characters which can be easily identified and simply inherited.
3. Through the maintenance of pedigree record the breeder may be able to obtain the information about inheritance of characters.
4. Plants or progenies with weaker and visible defects are eliminated at an early stage in the breeding programme.

5. This method gives maximum importance on the breeder to use his/her skill and judgement about the selection of plants and progenies.
6. This method takes less time than bulk method to release a new variety.

Demerits

1. The success of the method is mainly dependent on the skill of the breeder.
2. To keep the individual pedigree record is laborious and time consuming, it may be the limiting factor for large breeding programme.
3. Selection of large number of progenies in every generation is also laborious and time consuming.
4. In F_2 and F_3, the selection for yield is not effective. If sufficient number of progenies is not retained, valuable genotypes may be lost in early segregating generations.

Achievements

Pedigree method is useful in selection of new superior recombinant types from a hybridization programme. This method is suitable for improving specific characteristics, such as disease resistance, plant height, maturity time, etc. as well as yield and quality characters.

Many improved varieties have been developed through pedigree method in many crops like wheat, rice, barley, pulses, oil seeds, cotton, tobacco, jowar, vegetables, etc.

Wheat

K 65 (tall variety) ← C 591 x NP 773

K 68 (good quality grain) ← NP 773 x K 13

WL 711 (dwarf, high yield) ← (S308 x Chris) x Kalyan Sona

Malviya 12 (good grain) ← NP 876 x Cno 66

Rice

'Jaya' and 'Padma' (short duration, finer grain) ← Taichung Native 1 x T 141.

Cotton

'Laxmi' (fibre quality, early maturing, resistant to leaf blight) Gadag 1 x CC 2 (Cambodia Coimbatore 2).

Tomato

Pusa early dwarf (more yield) ← Meerut x Red cloud.

Bulk Method

After hybridization programme when the F_2 and subsequent generations are harvested in mass or as bulk to raise the next generation, it is called bulk method. At the end of bulking period, individual plants are selected and evaluated in the same manner of pedigree method. This method is applicable to any self-pollinating crop like cereals, legumes and oil seeds.

Three different advantages may be achieved from this method:

1. Isolation of homozygous lines,
2. Waiting for selection by environmental disaster,
3. The long period of bulking may be helpful for natural selection to change the composition of population.

Procedure

First Year

Hybridization is done between the selected parents with the desirable attributes. The cross may be simple or complex depending upon the number of selected parents.

Second Year

The F_1 generation seeds are space planted and seeds from these plants are harvested in bulk.

Third Year to Seventh Year

Starting from F_2 generation onwards the seeds are planted at commercial seed rate and spacing. The seeds of subsequent generations are harvested in bulk. Many environmental factors, disease outbreak, etc., may select out the particular genotype from this bulk population. The population size in each generation should be large always, 30,000 – 50,000 plants in each generation.

Eighth Year

F_7 generation seeds are space planted and superior phenotypes are selected and harvested separately. The selection is made on disease reaction, grain character, etc.

Ninth Year

Individual plant progenies of F_8 generation are grown in single or multi-row plots. Weak or inferior progenies are rejected, only 100-300 plant progenies are selected which are superior as well as showing no segregation.

Tenth Year

Preliminary yield trial is conducted with F_9 generation with standard commercial variety as check. The undesirable progenies are rejected.

Eleventh to Fifteenth Year

Replicated yield trial at several locations is carried out using standard commercial variety. The lines are evaluated for some important characteristics in addition to yield, disease resistance, etc.; superior lines are released as new variety.

Sixteenth Year

The seeds are multiplied and distributed to the farmer.

Schematic Representation

Year	
First Year	Selected parents are hybridized.
Second Year	F_1 Plants space planted, seeds are harvested in bulk.
Third Year	F_2 Plants are planted at commercial seed rate, seed harvested in bulk.
Fourth Year	F_3 ↓
Fifth Year	F_4 ↓
Sixth Year	F_5 ↓ For more five-six generations the same method is followed.
Seventh Year	F_6 ↓
Eighth Year	F_7 is space planted and individual plants are selected and seeds are harvested separately.
Ninth Year	Individual plant progenies are grown. Inferior progenies are eliminated.
Tenth Year	Preliminary yield trials using standard varieties as check.
Eleventh to Fifteenth Year	Multilocation yield trials for the selected progenies.
Sixteenth Year	Multiplication of seed for distribution.

Merits

1. This method is simple, convenient and less expensive.
2. The natural disease epiphytotics, winter killing, etc., eliminate the undesirable types and

increase the frequency of desirable types, which is more helpful for isolation of desirable types.

3. As natural selection acts here, so the progenies selected from long term bulk are far superior than those selected from F_2 of short term bulk.
4. Less attention and labour is needed for this method.
5. Since large populations are grown, the transgressive segregants are more likely to appear and there is greater chance of isolation of transgressive segregants in this method.
6. Artificial selection may be practiced to increase the frequency of desirable types.
7. Survivability of any gene or a particular genotype can be studied in this method.

Demerits

1. This method takes a long time to release a new variety, as the method of natural selection becomes operative only after F_8 or F_{10}.
2. The short term bulk method is useful for isolation of homozygous lines but the effect of natural selection has little effect on the genetic composition of the population.
3. Breeder has little opportunity to show his skill in this method, practically selection is mainly based on natural selection.
4. Inheritance of characters is not individually maintained as in pedigree method.
5. At the end of bulking period a large number of progenies are being selected.

Achievements

This method has been used to a limited extent due to lack of popularity. In U.S.A the barley varieties like Arivat, Beecher, Glacier and Gem have been originated from a bulk population derived from cross Atlas x Vaughn, the bulk was maintained in this case for 7 generations.

As this method requires more time to get the effect of natural selection for developing a new variety and the lack of using any skill of breeder for selection of superior types, this method is less used by the breeder for crop improvement.

Back-cross Method

Back-cross is the event of crossing of F_1 with either of the parents, but here in the back-cross method of breeding the hybrid is to be crossed with the superior parent whose genotype is to be transferred to the local variety. For example, in a particular area there is a well-adapted high yielding local variety but susceptible to a particular disease, this variety will be considered as recipient parent.

The disease resistant variety will be the donor parent. The recipient parent will be used in this method of breeding repeatedly to get all the genes to be transferred which is also known as recurrent parent and the donor parent is known as non-recurrent parent.

Genetic Consequences of Repeated Back-crossing

Repeated back-crossing results in decreasing the frequency of heterozygosity by 50% in each subsequent generation and rapid increase in homozygote, upto 98% by sixth back cross progeny, will be achieved. The genotype of the back-cross progeny becomes increasingly similar to that of the recurrent parent.

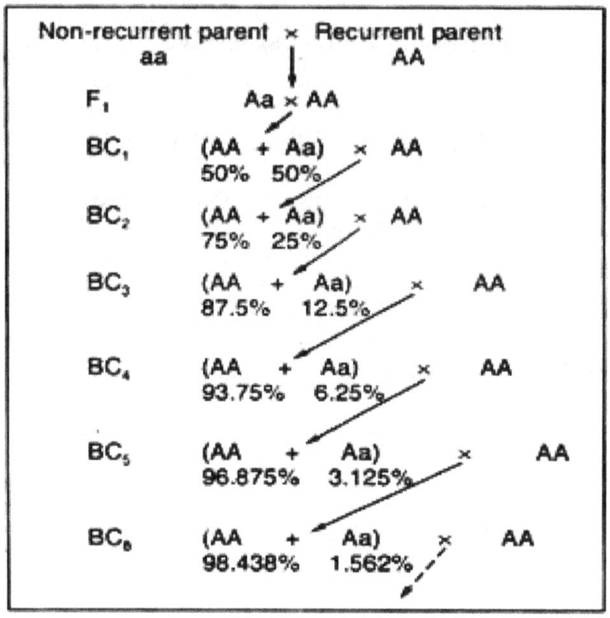

Procedure

The plan of back-cross method differs on the gene which is to be transferred, whether it is dominant or recessive.

(a) Transfer of Dominant Gene

Selection of Parents

Variety A — Recurrent parent (well adapted, high yielding); Variety B — Non-recurrent parent, dominant gene controlled (stem rust resistant).

- Hybridization (First Year): The recurrent parent (variety A) should be used as female parent and the non-recurrent parent (variety B) as male parent.
- F_1 Generation (Second Year): F_1 plants are back-crossed to variety A, in this generation all the plants are heterozygous for rust resistant character.
- BC_1 - Generation (Third Year): 50% of the BC1 plants which are resistant are selected and back-crossed to the recurrent parent, i.e., variety A.
- BC_2 - BC_5 Generation (Fourth-Seventh Year): In each back-cross generation, segregation would occur for rust resistance. Rust resistant plants are selected and back-crossed to recurrent parent, variety A.

- BC_6 - Generation (Eighth Year): The plants of this generation will have 98.4% genes from variety A. Rust resistant plants are selected and selfed. Their seeds are harvested separately.
- BC_6 - F_2 Generation (Ninth Year): Individual plant progenies of selfed plants are grown separately. The rust resistant plants are selected and the seeds are harvested separately.
- BC_6 - F_3 Generation (Tenth Year): Individual plant progenies are grown. The plants homozygous for rust resistance and similar to the plant type variety A are harvested in bulk.

Yield Test (Eleventh Year)

The new variety is tested in replicated yield trial along with the variety A as a check, as the new variety will be same in performance with variety A only with rust resistant character.

Twelfth Year

Seed multiplication is done and distributed to the farmers.

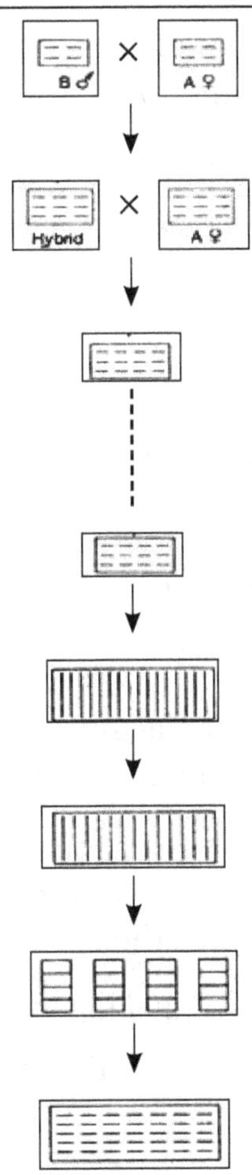

First Year	Hybridisation between variety A and B. A is the recurrent parent, B is non-recurrent.
Second Year	The hybrid plant is back-crossed with recurrent parent A
Third Year	The plants are selected which are rust resistance as well as similar to recurrent parent A. BC_2-BC_5 — Same procedure is followed for every year.
Eighth Year	BC_6 – Rust resistant plants are self pollinated. The seeds are harvested separately.
Ninth Year	Individual plant progenies are grown. The rust resistant plants similar to plant type A are selected.
Tenth Year	Individual plant progenies are grown. Homozygous progenies similar
Eleventh Year	Replicated yield trial with recurrent parent as one of the check.
Twelfth Year	Seed multiplication and distribution.

(b) Transfer of Recessive Gene

Procedure for transferring a recessive gene will be different from that of dominant gene, as the recessive gene will be expressed only in homozygous condition. So the selection for that recessive gene requires the F_2 generation, i.e., selfing is needed after every two back-crosses, and testing for the presence of that character has to be done in that generation.

Selection of Parents

The parents are selected in this programme as the previous one, only difference is that the rust resistance is a recessive gene controlled character:

- Hybridization (First Year): The recurrent parent is used as female parent and hybridized with the rust resistant donor plant.
- F_1 Generation (Second Year): The F_1 plants won't be rust resistant, these will be back-crossed to the recurrent parent.
- BC_1 Generation (Third Year): Here also no rust resistant plants will appear so no need to do any rust susceptibility test. All the plants are selfed.
- BC_1 - F_2 Generation (Fourth Year): The plants are tested for rust resistance character. These plants are selected and back-crossed with recurrent parent.
- BC_2 Generation (Fifth Year): There is no rust resistance test, the plants are selected only on resemblances with recurrent parent A, and again back-crossed with recurrent parent.
- BC_3 Generation (Sixth Year): Again the plants are self-pollinated, but without any selection for resistance character.
- BC_3 - F_2 Generation (Seventh Year): The selection for rust resistance is done in this generation. The rust resistant plants having resemblance with variety A are selected and back-crossed.
- BC_4 Generation (Eighth Year): Same as BC_2 generation.
- BC_5 Generation (Ninth Year): Same as BC_3 generation.
- BC_5 - F_2 Generation (Tenth Year): Plants are selected against rust resistance and also the characters of variety A. The selfed seeds from the selected plants are harvested separately.
- BC_5 - F_3 Generation (Eleventh Year): Individual plant progenies are grown and subjected to rust resistance test. Rigid selection is done for resistance to stem rust and the characteristic of variety A. Seeds from selected plants are mixed together to constitute the new variety.
- Yield Test (Twelfth Year): The variety is tested in replicated yield trial along with the variety A as check. The new variety should be same in yield performance except bearing the new character introduced.
- Thirteenth Year: Seed multiplication and distribution to the farmers are carried out.

First Year	B ♀ A ♂	Hybridisation between variety A and B. A is the recurrent parent, B is with recessive rust resistance gene.
Second Year	Hybrid A	F_1 is back-crossed with recurrent parent. No rust resistance test.
Third Year	BC_1	No rust resistance test. The plants are all self pollinated.
Fourth Year	$BC_1 - F_2$ × A	The rust resistant plants are selected and back-crossed with recurrent parent.
Fifth Year	BC_2 × A	No rust resistance test. The plants are back-crossed with recurrent parent.
Sixth Year	BC_3	As in BC_1.
Seventh Year	$BC_3 - F_2$ × A	As in BC_1-F_2.
Eighth Year	BC_4 × A	As in BC_2.
Ninth Year	BC_5	No rust resistance test. The plants are self pollinated to raise BC_5-F_2 generation.

Comparison between the transfer of dominant and recessive gene during back-cross method:

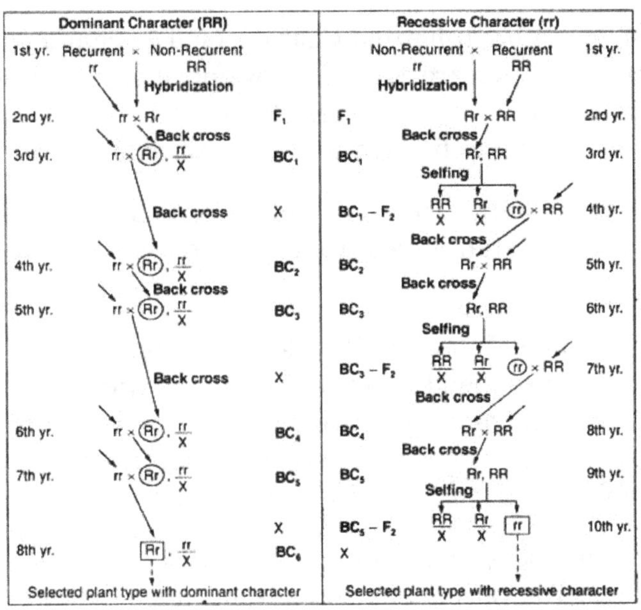

(c) Transfer of Quantitative Character

Quantitative characters are governed by polygenes such as grain size, plant height, maturity time, etc. As during transfer intensity of characters will be lost, so the non-recurrent parent must have the characters more intense from than it is desired in the new variety. For transferring this type of character, each back-cross generation should be selfed and F2 generation is grown which should have large population.

Then the rigid selection is done for the character to be transferred. The system of alternate back-crossing and selfing goes on till the desired results are obtained. At the end of back-cross programme, F2 is handled like pedigree method. The progenies which are more likely to recurrent parent bearing the desirable trait are selected and mixed to make up the new variety.

(d) Transfer of Two or More Characters in a Single Parent

For simultaneous transfer of many characters in a single back-crossing programme, the characters from different non-recurrent parents are brought together by multiple crossing and then back-crossing to recurrent parent is performed. Larger back-cross progenies would be needed than in case of transfer of a single character, and also it may take long time as in each step all the desired characters may not get expressed.

The multiple characters can be transferred in a stepwise method. The recurrent parent is first improved for one character then it is used as recurrent parent for introduction of another new character. Thus in stepwise manner more than one characters can be introduced into a single recurrent parent.

Applications

Generally the.back-cross method is used for transferring disease resistance character to a good and well adapted variety. Other quantitative characters can also be transferred through back-cross method in both cross- and self-pollinated crop.

1. Inter-varietal Transfer of Simply Inherited Characters: The characters like disease resistance, seed colour, plant height, etc., which are controlled by one or two major genes are most suited for transfer through back-cross method from one variety to another of the same species. The successful transfer depends on the minimum linkage between desirable and undesirable trait.

2. Inter-varietal Transfer of Quantitative Characters: Grain characters like seed size and shape, earliness, plant height-all can be transferred from one variety to another with the criteria of high heritability.

3. Interspecific Transfer of Simply Inherited Characters: Mainly the character like disease resistance can be transferred from related species to cultivated species. This specially requires the relatedness between the species where the chromosomes can pair during meiosis. The transfer will be unsuccessful if the genetic environment of recurrent parent is not suitable for functioning of the gene of desirable character from donor parent.

4. Transfer of Cytoplasm: In case of transfer of male sterile character from one parent to another requires back-cross method. The variety or species from which the cytoplasm is to be transferred

is used as female parent. The recurrent parent should be the male parent. After 6-8 back-crosses, the cytoplasm will be of the donor parent with the genotype of recurrent parent.

5. Transgressive Segregation: Modification of back-cross method will produce transgressive segregants. Few (1 to 3) back-crosses with F1 allow much heterozygosity to appear, or two or more recurrent parents may be used in back- cross programme to accumulate genes from them into the back-cross progeny. Such kind of modification would produce new variety which won't be like the recurrent parent.

6. Production of Isogenic Lines: Back-cross method is useful for production of isogenic lines, i.e., the lines of a crop which are identical in their genotype except for one gene. These isogenic lines are useful for studying the effect of individual genes.

Merits

1. This method does not change the genotype of the popular established variety, only it helps a single desirable character to be transferred in the existing variety.

2. As the recurrent parent is an established variety, so it is not necessary to test the yield performance which ultimately saves five years-time, as well as expenses.

3. In case of short duration plant, 2-3 generations can be raised within a year as the selection is based on inheritance of a particular character, not on performance. So this method drastically reduces the time period required to develop a new variety.

4. Back-cross method requires smaller populations than the pedigree method.

5. Defects of an established variety can be removed using this method, may be only by introducing a single character.

6. Interspecific gene transfer only can be done through this method.

7. Transgressive segregation can be obtained in case of quantitative characters by modified back-cross method.

8. This method is very much useful for cytoplasmic gene transfer to the recurrent parent, i.e., new variety will bear the cytoplasm of donor parent and genotype of recurrent parent.

Demerits

1. The new variety is not superior in performance than the existing variety except the introduction of single character.

2. During transfer of such kind of single desirable gene, sometimes some undesirable genes may get transferred.

3. For introduction of more than one gene controlled character, multiple crossing programme is necessary which is often difficult, time taking and costly.

4. Once a recurrent parent taken in a back-cross method may get replaced by another superior variety of high yielding ability.

Achievements

Back-cross method is very much useful for transferring of simply heritable character like disease resistance to well adapted popular local variety.

In case of wheat, 'Kalyan Sona' is the popular established variety to which the leaf rust resistance character has been transferred from diverse sources like Robin, KI, Bluebird, Tobari, etc. using back-cross method.

In case of Bajra, Tift 23A, a male sterile line which was susceptible to downy mildew has made resistant through back-cross method.

For interspecific transfer of genes, back-cross method is widely used for crop improvement. Cultivated sugarcane (Saccharum officincirum) is susceptible to pests and disease, crossed with S. spontanium which is resistant. This brings resistance but with undesirable characters like more fibre, low sugar, thin stem, etc. By back-crossing with noble cane these undesired characters are removed.

In cotton, the hybridization between Gossypium hirsutum and G. arboreum yielded highly sterile F hybrid, few tetraploid seeds were obtained. These plants were then back- crossed with G. hirsutum and two varieties have been selected from the back-cross progeny which are now being widely cultivated in Gujarat.

Except the characters of disease resistance other characters can also be transferred by the back-cross method. Such as in cotton, the ginning quality of fibre has been improved by back-cross method.

Breeding Methods for Cross-Pollinated Crops

The breeding methods used for cross-pollinated crops are In this method of breeding, the best individuals with desired characters are selected on the basis of phenotypic performance in a source population. Open-pollinated seeds of the selected individual plants are divided into two halves. Second year replicated progeny row trial is conducted using one set of half seeds from each plant.

On the basis of the progeny performance, the best parental individuals are identified. The remnant half seeds from the superior parental plants are mixed and grown in isolation for random mating during the third year.

This method of breeding is equivalent to ear-to-row selection in context of maize originally proposed by C.G. Hopkins at the Illinois Agricultural Experiment Station in 1896 to improve protein and oil content of maize. This method has been named as mass-pedigree method by S.S. Rajan in India. This very method is called line breeding when selection is based on progeny tests and a group of progeny lines is composited.

Inbreeding

The mating of individuals more closely related than individuals mating at random is known as inbreeding. The lines produced by continued inbreeding are known as inbred lines. Self-fertilization is the most intense form of inbreeding.

In plant breeding nearly homozygous lines are produced by continued self-fertilization accompanied by selection for five to six generations. This can be used as the method of breeding only in those crops, which do not show any loss of vigour due to inbreeding, like cucurbits.

The three important uses of inbreeding in cross-pollinated crops are as follows:

- To attain uniformity in plant characters.
- To improve yield etc. by individual plant selection as in cucurbits in which there is no inbreeding depression.
- To develop suitable inbred lines in production of hybrids and synthetics.

Synthetic Variety

The term 'synthetic variety' has come to be used to designate a variety that is maintained from open pollinated seed following its synthesis by hybridization in all combinations among a number of selected genotypes, which have been tested for combining ability.

The components of a synthetic variety could be inbred (usually), clones, mass selected populations or various other materials. The component units are maintained so that the synthetic may be reconstituted at regular intervals.

The inbreeds to be used as component lines are chosen on the basis of combining ability tests. The component inbred are crossed in all possible combinations. This inter-crossed seed is called as Syn 0.

Equal quantity of seed from all crosses is composited and the mixture is allowed open-pollination in isolation and seed is harvested. This becomes Syn 1 generation. In absence of reconstitution of a synthetic at regular intervals, the population becomes an open-pollinated variety.

The testing for combing ability is the decisive criterion for a synthetic variety by which it can be distinguished from a conventional variety of a cross-pollinating species, which originates in a continuous selection of individuals and subsequent progeny tests. The greater variability caused by crossing several components with high general combining ability makes the synthetic varieties more adaptable compared to conventional varieties.

Similar to hybrids, the yield of a synthetic variety generally also decreases after the Syn 2, until an equilibrium is reached which, in partially self-fertile species, depends on selfing rate and inbreeding (minimum depression), but also on the number of components used in the Syn 0. The Performance of Syn 1 can be Estimated by the formula:

$$\hat{F}_2 = \bar{F}_1 - \left(\frac{\bar{F}_1 - \bar{P}}{n} \right) 8$$

The yield to be expected, usually increases with the number of components until an optimum is reached. The question regarding the most favourable number of genotypes the Syn 0 should be composed of cannot be clearly answered, because the evidence from research and practice is too divergent. The yield can be increased by:

1. An increase of the mean performance of the F1 combinations.
2. An increase of the mean performance of the parents (inbred lines, clones).
3. An optimum combination of the components.

Therefore, it is obvious that I_0 plants should already be tested for their combining ability and plants or lines should be used as components, the inbreeding depression of which is not as strong as in the I_5. In order to maintain performance in subsequent generations, mass selections have been found to be sufficient in maize.

The performance of synthetics can be improved by one further breeding cycle. It consists of selection of genotypes from the synthetic variety, their testing by a dialled, and combination of genotypes with the highest combining ability for a new synthesis (= recurrent selection).

Composite Varieties

Concept of composite varieties (in maize) originated in India. Composite varieties are generally derived from the varietal crosses in advanced generation. These are usually developed from open-pollinated varieties or other heterozygous populations or germplasm which have originally not been subjected to inbreeding or have not been elaborately tested for their combining ability.

Usually, they involve open pollinated varieties, synthetics, double crosses, etc., selected for yield performance, maturity, resistance to diseases and pests. These composites often show a high order heterosis in F_1's when widely diversed populations are crossed. Advanced generations of such heterotic crosses often show stabilized yields. General combining ability and additive gene effects play predominant role in exploitation of these populations.

The details of the steps involved in development of composite variety are as follows:

- Screening of diverse germplasm by evaluation at multi-locations/years to identify the sources having adaptability, desirable agronomic attributes and resistance to major diseases and tolerance/resistance to serious insects.
- Making of all possible crosses among selected superior genotypes or top crossing with a varietal complex of screened base varieties.
- Conducting multi-location test with the F_1 and F_2 generations of varietal crosses and selection of F_2 's showing desirable agronomic features along with least decline in F_2.
- Evaluation of selected F_2 populations and identification of the best one as practically composites are constituted by compositing seeds of various populations and allowing the mixture to stabilize under open pollination along with some mild selection in isolation. The constituent entries may not be maintained for reconstituting the composite. Composite may serve as a base population for developing inbred lines.

Hybrids

The term hybrid variety is used to designate F_1 populations that are used to commercial planting. The F_1 's are obtained by crossing genetically unlike parents. The pioneering work on hybrid maize was done by G.H. Shull (single crosses) and D.F. Jones (double crosses). Other innovative researchers in this area have been E.M. East, H.K. Hayes, F.D. Richey and M.T. Jenkins and others.

Types of Hybrids

(i) Single cross: A single cross is a hybrid progeny from a cross between two unrelated inbred.

$$= A \times B$$

(ii) Three-way cross: A three-way cross is the hybrid progeny from a cross between a single cross and an inbred.

$$(A \times B) \times C$$

(iii) Double cross: A double cross is the hybrid progeny from a cross between two single crosses.

$$(A \times B) \times (C \times D)$$

(iv) Modified single cross: A modified single cross is the hybrid progeny from a three- way cross which utilizes the progeny from two related inbred as the seed parent and an unrelated inbred as the pollen parent.

$$(A \times A') \times B$$

(v) Double modified single cross: A double modified single cross is the hybrid progeny from two single crosses, each developed by crossing two related inbred.

$$= (A \times A') \times (B \times B')$$

(vi) Modified three-way hybrid: A modified three-way hybrid is the progeny of a single cross as female parent and another single cross between two related inbred.

$$= (A \times B) \times (C \times C')$$

(vii) Top cross hybrid: This is inbred x variety hybrid. Following top cross hybrids may be formed:

(a) Inbred line x variety

(b) Inbred line x experimental hybrid

(c) Inbred line x synthetic variety

(d) Inbred line x family

(viii) Double top cross hybrid: A double top cross hybrid is the progeny of a single cross and a variety. Such hybrids have been produced on commercial scale in India and China.

Methods of Inbred Line Development

(i) Standard selfing method: Self-Pollination of individual plants within single plant progenies grown is the most common procedure used to develop inbred lines. This breeding procedure has two important problems.

(a) Vigour of the lines is decreased with inbreeding because of loss of favourable dominant allelles and any heterozygous loci that have over-dominant effects. Many lines are so poor in seed yield, pollen production, etc., that they cannot be used in a programme to produce single cross hybrid seed.

(b) Effective selection within the row for the plants that have desired agronomic traits becomes minimal in generations beyond S3. Therefore, some breeders use only two or three generations of self-pollination with subsequent reproduction by sib-mating within progenies.

(ii) Special techniques: These techniques involve the doubling of haploids derived from either maternal or paternal gametes. Somaclonal variations from inbred lines offer another opportunity. However, due to low frequency of haploids and doubling of haploids to the diploid state, these methods are still not an important component of most breeding programmes.

Besides, these two methods, there are three other methods where along with developing inbred lines, there are opportunities to improve them simultaneously.

(iii) Pedigree method: In this method, a pair of elite lines that complement one another are crossed to produce the F_2 generation and pedigree selection is practiced by sampling the F_2 population. F_2 populations of the single crosses are the most frequently used source populations for line development.

(iv) Backcrossing: This is a modification of pedigree method. Modifications of backcross method have also been suggested for example, in convergent improvement by Richey, there is parallel improvement of two inbred lines by the reciprocal addition of dominant favourable genes present in one line and lacking in the other line.

In this method, two inbred A and B are crossed. The F_1 is backcrossed with A followed by selection of desirable traits of B and F_1 is also backcrossed with B where selection for desirable traits of A is made. After about three backcrosses and selection, selfing is done to fix the selected genes. This method is useful for improving such characters as vigour, resistance to diseases, pests and lodging.

(v) Gamete selection: This scheme devised by L.J. Stadler in 1944, is based on the premise that if superior zygotes occur with a frequency of p2, superior gametes would occur with a frequency of p. The procedure involves crossing an elite line with a random sample of pollen of plants from a source population.

The resulting F_1 plants and the elite line are testcrossed to a common tester and F_1 plants are also selfed. Testcross progenies are evaluated in a replicated trial. The test crosses of F_1 plants that exceed the elite line by tester are presumed to have obtained superior gametes from the source population. Superior gametes are recovered as F_2 self's.

Selection and selfing are continued till desirable homozygosity/uniformity is attained. Though gamete selection is not used as extensively as pedigree and backcross methods, it does have some intrinsic features, and consequently, is included in some breeding programmes.

Testing of Inbred

The general combining ability of the inbred is tested by making all possible crosses $\left[n\,(n-1)/2\right]$ in a diallel fashion or else top cross test is carried out. Specific combining ability is also estimated.

Time of Testing

One procedure is to test the inbred for hybrid performance/combining ability in about the fifth generation of selfing when the number of selected lines is greatly reduced. This breeding system assumes favourable relationships of plant and other traits of inbred lines (traits as selection criteria during inbreeding) with combining ability for grain yield.

A second system of inbred development is based on an evaluation for hybrid performance in the early generation of self- pollination, e.g., testcrosses of the So plants or S1 lines. Genotypes that are identified for above average performance in these tests are continued in the selfing and selection nursery.

This procedure has been called as 'early testing' originally proposed by M.T. Jenkins in 1935. The assumption is that the combining ability of a line is determined early in its development and will change relatively little in subsequent generations of inbreeding and selection.

By early testing, the breeder is in a position to discard some lines that are inadequate in hybrid performance and wasteful expenditure on these lines is avoided. However, probably most breeders use a method that is intermediate between these two systems. In this approach, first hybrid evaluation are of S_2 or S_3 lines.

Combination of Inbred in Hybrids and Prediction of Double Cross Performance:

Single Crosses: $n\,(n-1)\,/\,2$

Three-way crosses: $n\,(n-1)\,(n-2)\,/\,2$

Double crosses: $n\,(n-1)\,(n-2)\,(n-3)\,/\,8$

where, n = Number of inbred

If there are 4 inbred, A, B, C and D, the performance of the double cross (AxB) (CxD) is predicted as follows:

(i) Based on mean performance of all possible six single crosses:

$$(A\times B)+(A\times C)+(A\times D)+(B\times C)+(B\times D)+(C\times D)/6$$

(ii) Based on mean performance of the four non-parental single crosses:

$$(A\times C)+(A\times D)+(B\times C)+(B\times D)/4$$

(iii) Based on mean performance of four lines over a series of single crosses:

$$(A\times E)+(A\times F)+(A\times G)+(A\times H)+(B\times E)+(B\times F)+(B\times G)+(B\times H)+(D\times H)/n$$

(iv) Based on mean performance of top-crosses of the four inbreeds

$$(A \text{ x variety}) + (B \text{ x variety}) + (C \text{ x variety}) + D \text{ x variety}/4$$

Of the above four methods, method (ii) is found to give more accurate results.

Genetical Basis of Heterosis

The phenomenon of hybrid vigour, expressed particularly in the first generation (F_1) following the crossing of cultivars or inbred lines, has been known for more than a hundred years. The term heterosis, coined by G.H. Shull in 1909 suggests a mechanism based on heterozygosity and therefore, is not fixable in the homozygous state. Two hypotheses have been put forward to explain heterosis.

(i) Over-dominance Theory: This was proposed by G.H. Shull and E.M. East independently in 1908. According to this hypothesis, hybridity/heterozygosity is superior to either homozygote and this state of heterozygosity has a stimulating effect upon the physiological activities of the organism leading to superiority of Aa over AA or aa.

(ii) Dominance Theory: This hypothesis was proposed by C.B. Davenport in 1908, A.B. Bruce in 1910 and F. Keeble and C. Pellew in 1910. According to this hypothesis, each dominant allele contributes equally to heterosis and the recessive alleles contribute nothing. It is also assumed that the dominance is complete.

If all the dominant alleles are concentrated in one parent, and the counterpart recessive alleles in another parent, the F_1 will be equal to the parent having all the dominant alleles.

AABBCCD × aabbccd

P1 ↓ P_2

F1 AaBbCcd

In the above cross, if each dominant allele, contributes 1 unit and the recessive allele, 0 unit, then the P_1 will have a value of 4 and P_2 a value of 0. F_1 will have a value of 4. However, there are situations, where F_1 is superior over the better parent.

These cases under dominance theory can be explained assuming that dominance and recessive alleles are distributed in both the parents as given below:

AABBCCdd × aabbccd

P1 ↓ P_2

F1 AaBbCcDd

In this cross, the phenotypic value of P_1 is 3, that of P_2 is 1 and F_1 has a score of 4 which is superior to the better parent. Under this model it should be possible to derive a pure line from the F1 which should be equal to F_1 in the performance and thus this heterosis will be fixable.

However, it is generally agreed that heterosis is not fixable in the homozygous state. This would be the case if hybrid vigour were due to true over-dominance or due to tight linkage in the repulsion phase at some incompletely dominant loci.

Much evidences suggest that apparent over-dominance is, in fact, due to non-allelic interaction and linkage disequilibrium and that heterosis is mainly a result of the bringing together of unidirectionally dominant alleles distributed between the parental line. Under this, heterozygosity is not an essential prerequisite for high performance, uniformity and stability of performance.

Recurrent Selection

Recurrent selection is a method of breeding designed to concentrate favourable genes scattered among a number of individuals by selecting in each generation among progeny produced by matings inter-se of the selected individuals (or their selfed progeny) of the previous generation.

Based on the ways in which plants with desirable characters are identified, recurrent selection has been divided into four types.

- Simple recurrent selection or recurrent selection for phenotype
- Recurrent selection for general combining ability
- Recurrent selection for specific combining ability
- Reciprocal recurrent selection

In simple recurrent selection a number of plants are self-pollinated in a source population in first year. At maturity superior plants based on phenotypic performance are selected. In second year, seeds produced by self-fertilization of the selected plants are planted and crossed in all possible combinations and the produce is bulked.

This completes original selection cycle. Since selection is based on the phenotype of the plant, it is useful only for characters with high heritability. In those cases, where it is possible to identify the desired selections before flowering as in case of cauliflower, cabbage, etc., inter-crosses of selections may be made in the first year of each cycle and the second year may be eliminated from each cycle.

Thus, strictly speaking, selfing is not an integral component of simple recurrent selection, rather it is done only to prevent crossing from the inferior pollen grains before the plants reach to selection stage.

In recurrent selection for general combining ability, a three year cycle is involved. In first year a number of plants are self-pollinated and crossed to a broad based heterozygous tester stock to identify the So plants with good general combining ability. In second year, the crosses are evaluated to identify those that are superior. Self's of first year are kept in reserve.

In third year, the reserve selfed seeds are grown out, inter-crossed in all combinations, and a composite of inter-crossed seeds is used to establish an improved population for further selection. This procedure developed as a direct outgrowth of studies of early testing first proposed by M.T. Jenkins in 1935.

Recurrent selection for specific combining ability was proposed by F.H. Hull in 1945. This method of selection is same as that of recurrent selection for general combining ability except that the tester selected is a narrow base an inbred line. The recurrent selection for general and specific combining ability is equivalent to half sib progeny test.

Reciprocal recurrent selection proposed by R.E. Comstock, H.F. Robinson and P.H. Harvey in 1949 aims at simultaneous improvement of two heterozygous and heterogenous populations (designated as A and B).

A serves as tester for B and B serves as tester for A. This method is as effective as recurrent selection for gca when additive gene action predominates, and is as effective as recurrent selection for sea when non-additive effects are of major importance. The Steps are as follows:

I. Season: Selected plants of population A are self-pollinated and crossed to plants of population B. Likewise plants are selected and self-pollinated in B and outcrossed to plants of population A.

II. Season: Test cross progenies of both the populations are evaluated in replicated trial. Superior progenies are identified on the basis of performance in this trial.

III. Season: Selfed seed from plants with superior test cross progenies are grown population wise separately and inter-crossed to reconstitute two populations which will be now called as A' and B'. This completes one cycle and additional cycle (s) may be initiated.

Breeding for Stress Management

Drought Resistant Crops

Adverse environmental factors are examples of the most severe constraints to agriculture not only to developing countries but to the world at large. Water scarcity being one of the factors contributes to the highest percentage death rate of different crops. Climate changes together with increasing population are some of the factors which contribute to water scarcity and thus the importance of water use efficiency (WUE) improvement in agriculture. This requires the development of crop varieties with increased tolerance to drought through conventional breeding and genetic engineering methods thus meeting global food demands with less water.

Drought resistance (DR) being a complex traits triggers a wide array of physiological responses in plants which affects the activity of various genes. Plants respond to the ever changing environment in a complex way that allows them to react to different conditions at a given time. Therefore, the genetic control of tolerance to abiotic stress is not only complex but also influenced by other environmental factors at various stages of plant growth.

Plants use multiple ways to respond to drought stress and have adapted to drought through physiological and morphological changes which are controlled by a cascade of diverse signals. The physiological responses of plant to drought include leaf wilting, leaf abscission, a reduction in leaf area and stimulation of root growth through directing nutrients to the underground plant parts (Alexandre Bosco). Plants also survive severe drought stress through a plant stress hormone called abscisic hormone (ABA) which induces closure of leaf stomata thus reducing water loss through transpiration. All these improve water use efficiency (WUE) of plants.

Plant response to drought is also accompanied by activation of genes which are involved in both perceptions of drought stress and transmission of the stress signals. Examples of these genes are those that govern the accumulation of compatibles solutes, stabilization of cell structures from desiccation by reactive oxygen species (ROS).Other genes regulate transduction of stress signal and modulate gene expression. All this gives an idea of potential improvement of DR of crops in field trials though the biochemical together with a molecular basis of drought perception, stress adaptation remains the major challenges for the genetic improvement of DR.

The Genetics of Drought Resistance

Genetic Control of Leaf Traits Related to Drought Resistance

Various leaf traits have been associated with DR mechanisms and these traits are highly used to evaluate DR of crops. This includes both morphological and physiological traits which are usually used as drought avoidance (DA) by plants. Plants also have drought tolerance (DT) mechanisms for osmotic adjustment and sugar content. All these together with other traits are among the factors considered by many breeders and researches who elucidate the genetic bases of leaf related DR traits by quantitative trait locus (QTL) mapping approaches. Little work has been done on improving crops for drought resistance through the integration of conventional breeding; marker assisted breeding (MAB) and genetic engineering.

Complex functional traits have been achieved by many researchers through QTL mapping approach due to the development of gene based markers and availability of functional information for various plants which ease functional annotation with QTL studies. Use of high throughput genotyping maps together with functional annotation have narrowed down the position of loci of the targeted traits thus identifying positional candidate genes for these QTL.

Detection of QTL involved in photosynthesis and WUE in the context of drought resistance is a mile towards understanding the genetic basis regulating the expression of these traits. Photosynthesis ability is mainly determined by stomata density and stomata aperture but genetic variation of these factors are unknown making it harder for researchers. Plants when exposed to water stress, drought or other harsh conditions have to cope with the stress or else die. The sensitivity and response time to drought differ from different species.Repeated drought encounter increases the sensitivity to environmental changes that induces stomata closure, while the sensitivity to changes that induce stomata opening is reduced.

Leaf rolling is another drought avoidance trait to overcome harsh environmental conditions. Leaf rolling reduces water loss in addition to reducing the leaf area exposed to heat and light radiation. Various plants differ to roll leaves under similar water deficit and there is evidence that enhanced ability to roll leaves confers a yield advantage under drought conditions.

Genetic Control of Physiological Traits Related to Drought Resistance

DT in crops has been associated with physiological traits like osmotic adjustment and cell membrane stability. Osmotic adjustment (OA) is defined as the lowering of osmotic potential due to the accumulation of solutes in response to water deficit. This accumulation of solutes has various functions related to OA which results to increased water retention and turgor maintenance. In some plants like peas (Pisum sativum L.), OA has been associated with increased yields under drought stress conditions.The overall effect of OA on plant production under stress conditions involves the whole plant development and this brings about deep root growth or higher root length density in the soil. QTL mapping analysis for OA under stress conditions has been done in various crops but care must be taken in mapping QTLs for OA and other physiological traits. Evaluation of genotypic difference in OA is supposed to be performed at the same level of stress in all genotypes because solute accumulation in a given genotype is a function of leaf desiccation.

Numerous field studies have been done in crops like wheat and Barley to examine the association of yields and water use efficiency (WUE) under drought stress conditions. Results indicated that plant utilizes different mechanism for moderating water use like reducing plant size and leaf area under stress conditions. An indirect indicator of WUE was difference in carbon isotope discrimination which assisted plant breeders in drawing a conclusion.

Plants also use ABA under stress condition as a DA thus playing a role in pre-invasive defense. Other than ABA's role in drought and low temperature stress tolerance ,ABA has also been reported to mediates plant defense responses.Despite the importance of ABA in DA ,its genetic basis of drought–induced ABA accumulation in relation to morphological traits like leaf size and root growth remains unknown.

Genetic Control of Root Traits Related to Drought Resistance

Plant roots play a vital role in the growth of a plant during stress conditions. Root hairs which are associated with maintaining plant productivity during drought conditions includes root qualities like diameters, length and density especially at depths in soil with available water.In dry areas, small xylem diameters in deep roots help in improving root acquisition of water when ample water at depth is available. Genetic control of root traits through breeding has been a bottleneck because of the complexity of the root system architecture (RSA).

Soil exploration increment by the root systems is a positive feature for improving drought tolerance and thus large genetic variability of root system characteristics is a source of genetic progress. In fields with deep soil and water table, roots grow deeper in the soil layers thus improving both water status and access to water because of hydraulic properties.

Multiple genes usually control various root traits where each governs small effects with a degree of interaction effects that can change with environmental conditions.The QTL that contribute to root traits can therefore be considered as either constitutive or adaptive and this makes it useful in selecting traits which are useful in the target environment.Both constitutive and adaptive methods are difficult to phenotype and this is one of the reasons why most genetic researchers focus on plants part above the soil as compared to the one below.

Root QTLs have shown a greater potential for marker assisted selection (MAS) mostly when chosen roots traits contribute to drought tolerance in the target environment. The selected function for use in MAS can vary depending on the environment at hand and the objective of the researcher. QTLs that have been identified in green houses conditions are supposed to be validated under field conditions and should relate to improvements in productivity before use in MAS program.

Improvement of productivity under drought conditions through utilization of molecular markers is a challenge in crop improvement simply due to root variation being difficult to phenotype.

Marker assisted selection in a breeding population of about hundreds of genotypes assist breeders with different options to select combinations of favorable traits above and below grounds. Molecular marker with strong linkage disequilibrium with the QTL of the desired root traits together with the gene are mostly identified and assist in plant breeding program for the success of MAS. Accurate phenotyping becomes the challenge of using MAS and this can result in greater QTL accuracy locations in higher density molecular maps.

Functional Genomics of Drought Resistance at Various Levels

Due to complex genetic basis and diverse mechanism of DR where thousands of genes are up or down regulated under drought stress conditions in specific organs, functional genomics seems to be a way out to study molecular basis of DR at the various levels which include proteomic, transcriptomic, metabolomic and epigenetic levels.

Transcriptomic analysis has mostly been used to identify drought responsive genes in various plant species.Transcriptome sequencing (RNA-seq) has improved the throughput of gene expression profiling in drought stressed plants. Microarrays have facilitated the identification of various drought resistant genes and reveal that the transcription level in response to drought is influenced by time of day and genotype.Common drought-responsive genes in different tissues and genotypes have specific functions in regulating stem development under drought. Using Affymetrix Gene Chip Popular Genome Arrays identified divergent responses in gene expression profiles in response to drought stress between two populous hybrids and conclude that it's not possible to describe a common genus-specific drought-driven Transcriptome based on few genotypes.

Proteome-wide profiling approach is another way used in characterizing plant phenotypes where the method targets different proteins and understand their role in stress. This makes it important to understand the complex biological mechanism which is involved in response to a biotic stress tolerance. Post-translational changes are also important where a single gene can translate into different proteins and few genes can lead to a diverse proteome. This explains the limitation to genomics and transcriptomic approaches more so when translational changes govern phenotype.

Epigenetic regulation is another way beside genetic control in which drought response can be regulated. This is through epigenetic mechanisms like DNA methylation and Histone modification. DNA methylation, plays a role in regulating genome wide gene expression and maintaining genomic stability in response to drought stress.Genome Wide methylation in Arabidopsis were recently shown to increase in the progeny of plants exposed to extreme temperature, UV-B, flooding and salt stress but decease in the progeny of drought stressed plants.

In Histone modification, post translation modification of specific residues on the core histone (H2A,H2B,H3 and H4) play a role in regulating genome functions by recruiting of transcriptional factors or other proteins that affect chromatic structure and state. Upon water deficiency, nucleosome pattern and histone modification specifically on the H3N-tail of four Arabidopsis drought responsive genes were altered to activate their coding regions.

Genetic Engineering of Drought-Resistance Genes in Crops

Through the studies of molecular, biochemical and genetics, drought stress signaling have been revealed to be regulated by transcriptional activation of drought induced genes, protein kinase, nitrogen –activated protein.

kinase cascades and other post translational processes. All these genetic regulations can be divided into transcriptional, post-transcriptional and post-translational aspects.

Transcriptional reprogramming plays a vital role in plant adaptation to drought and some of the transcriptional factors act as a central hub to modulate the expression of multiple-drought regulated genes.Transcriptional Factors (TFs) are example of Denovo synthesized regulatory proteins and in Arabidopsis Thaliana, more than 1700 genes encoding TFs in above 50 families have been identified where a good percentage of them have been associated with plant response to drought. Many transcription factors like basic leucine zipper (bZIP), ABA responsive element binding protein. NAM (no apical meristem), CUC2 (cup-shaped cotyledon), (NAC) like stress-responsive NAC and zinc-finger like C2H2 zinc finger protein (ZFP)) families have been characterized according to their roles in drought stress response regulation and expression of these TFs resulted in improved crop tolerance to drought.

In a post–transcriptional regulation, accumulation of small RNAs under drought has a role in stress related responses. In Arabidopsis, levels of 117 miRNAs have been analyzed using miRNA chips and 17 stress–inducible miRNAs have been identified Genome wide identification of drought responsive miRNAs have been carried in different plants like soybean, sugarcane with a number of conserved and non-conserved miRNAs related to drought response.

Post-translational modification plays a bigger role in signaling cascades within highly interconnected networks in response to biotic and abiotic stimuli. Protein Farnesylation is a post translational protein modification where farnesyl group to a target protein where it regulates many different growths and development processes. Farnesyltransferase consist of α- (FTA) and a β- (FTB) subunit where loss of function of FTB in Arabidopsis resulted in mutants that have an enhanced response to ABA.

Protein phosphorylation plays a vital role where some protein kinases need to be phosphorylated to become active. Various protein kinases have been associated with response to osmotic stress based on their transcriptional responses to different environmental stimuli.Some of the protein kinases are Mitogen Activated Protein Kinases(MAPKs),High Osmolarity glycerol (HOG1) etc. Some protein kinases involved in stress signaling are activated by calcium and they include CIPKs. This is through interaction with calcium binding proteins during the ABA response.

In plant cell, accumulation of reactive oxygen species (ROS) is mainly caused by abiotic stresses which results in damage to genetic material thus interfering with cellular energy homeostasis through activation of Poly(ADP-ribose) polymerase which catalyzes long branching of poly(ADP-ribose) polymers covalently attached to proteins by consuming NAD+ pools.

Breeding in Drought Resistance Crops

Conventional breeding and marker assisted selection (MAS) are examples of breeding techniques carried out in drought resistance crops. Conventional breeding for DR involves backcross strategy to develop new varieties with improved yield potential and DR. This breeding method requires the identification of genetic variability to drought among crop varieties where genetic variability traits are introduced into cultivars with the required characteristics. This method has been applied in breeding programs of various crops like rice, wheat, maize and one of its short backs is being a slow process which is limited by the availability of suitable genes for breeding and temporal variation in the field environment.

Marker assisted selection is a technique deployed to improve crop DR though traits like disease resistance. In this method, major QTLs for drought resistant donor genotypes are introduced into high yielding but fewer drought-resistant parents and thus the developed superior cultivars have only the major QTLs. Molecular markers like restriction fragment length polymorphism (RFLP), random amplified polymorphic DNA (RAPD) and isozyme facilitate development of drought resistant genotypes more effectively because their expressions are independent of environmental effects. This application of marker assisted selection in evolving drought resistant genotypes is still in progress and more experiments are being carried out.

Heat Stress in Plants

Plant breeding is process of development of new cultivars. Plant breeding involves development of varieties for different environmental conditions – some of them are not favorable. Among them, heat stress is one of such factor that reduces the production and quality significantly. So breeding against heat is a very important criterion for breeding for current as well as future environments produced by global climate change (e.g. global warming).

Breeding for Heat Stress Tolerance in Plants

Heat stress due to increased temperature is a very important problem globally. Occasional or prolonged high temperatures cause different morpho-anatomical, physiological and biochemical changes in plants. The ultimate effect is on plant growth as well as development and reduced yield and quality. Breeding for heat stress tolerance can be mitigated by breeding plant varieties that have improved levels of thermo-tolerance using different conventional or advanced genetic tools. Marker assisted selection techniques for breeding are highly useful. Recently 41 polymorphic SSR markers has been identified between a heat tolerant rice variety 'N22' and heat susceptible-high yielding variety 'Uma' for the development of new 'high yielding-heat tolerant' rice varieties.

What is Heat Stress Tolerance

Heat stress is defined as increased temperature level sufficient to cause irreversible damage to plant growth and development. Generally a temperature rise, above usually 10 to 15 °C above ambient, can be considered heat shock or heat stress. Heat tolerance is broadly defined as the ability of the plant tolerate heat – means that grow and produce economic yield under high temperatures.

Significance: Current and Future - Global Warming

Heat stress is a serious threat to crop production globally. Global warming is particularly consequence of increased level of green house gases such as CO_2, methane, chlorofluorocarbons and nitrous oxides. The Intergovernmental Panel on Climatic Change (IPCC) has predicted a rise of 0.3 °C per decade reaching to approximately 1 and 3 °C above the present value by 2025 and 2100 AD, respectively.

Physiological Consequence of Heat Stress

At very high temperatures cause severe cellular injury and cell death may occur within short time, thus leading to a catastrophic collapse of cellular organization. However, under moderately high temperatures, the injury can only occur after longer exposure to such a temperature however the plant efficiency can be severely affected. High temperature directly affect injuries such as protein denaturation and aggregation, and increased fluidity of membrane lipids. Other indirect or slower heat injuries involve inactivation of enzymes in chloroplast and mitochondria, protein degradation, inhibition of protein synthesis, and loss of membrane integrity. Heat stress associated injuries ultimately lead to starvation, inhibition of growth, reduced ion flux, production of toxic compounds and production of reactive oxygen species (ROS). Immediately after exposure to high temperature stress-related proteins are expressed as stress defense strategy of the cell. Expression of heat shock proteins (HSPs), protein with 10 to 200 kDa, is supposed to be involved in signal transduction during heat stress. In many species it has been demonstrated that HSPs results in improved physiological phenomena such as photosynthesis, assimilate partitioning, water and nutrient use efficiency, and membrane stability.

Studies have found tremendous variation within and between species, thus this will help to breed heat tolerance for future environment. Some of attempts to develop heat-tolerant genotypes are successful.

Traits Associated with Heat Stress Tolerance

Different physiological mechanisms may contribute to heat tolerance in the field—for example, heat tolerant metabolism as indicated by higher photosynthetic rates, stay-green, and membrane thermo-stability, or heat avoidance as indicated by canopy temperature depression. Several physiological and morphological traits have been evaluated for heat tolerance - Canopy temperature, leaf chlorophyll, stay green, leaf conductance, spike number, biomass, and flowering date.

(a) Canopy temperature depression (CTD): CTD has shown clear association with yield in warm environments shows it association with heat stress tolerance. CTD shows high genetic correlation with yield and high values of proportion of direct response to selection indicating that the trait is heritable and therefore amenable to early generation selection. Since an integrated CTD value can be measured almost instantaneously on scores of plants in a small breeding plot (thus reducing error normally associated with traits measured on individual plants), work has been conducted to evaluate its potential as an indirect selection criterion for genetic gains in yield. CTD is affected by many physiological factors, which makes it a powerful.

(b) Stomatal conductance: Canopy temperature depression is highly suitable for selecting physiologically superior lines in warm, low relative humidity environments where high evaporative demand leads to leaf cooling of up to 10 °C below ambient temperatures. This permits differences among genotypes to be detected relatively easily using infrared thermometry. However, such differences cannot be detected in high relative humidity environments because the effect of evaporative cooling of leaves is negligible. Nonetheless, leaves maintain their stomata open to permit the uptake of CO_2, and differences in the rate of CO_2 fixation may lead to differences in leaf conductance that can be measured using a porometer. Porometry can be used to screen individual plants. The heritability of stomatal conductance is reasonably high, with reported values typically in the range of 0.5 to 0.8.

Plants can be assessed for leaf conductance using a viscous flow porometer that is available on the market (Thermoline and CSIRO, Australia). This instrument can give a relative measure of stomatal conductance in a few seconds, making it possible to identify physiologically superior genotypes from within bulks.

(c) Membrane thermostability: Although resistance to high temperatures involves several complex tolerance and avoidance mechanisms, the membrane is thought to be a site of primary physiological injury by heat, and measurement of solute leakage from tissue can be used to estimate damage to membranes. Since membrane thermostability is reasonably heritable and shows high genetic correlation with yield.

(d) Chlorophyll fluorescence: Chlorophyll fluorescence, an indication of the fate of excitation energy in the photosynthetic apparatus, has been used indicator for heat stress tolerance.

(e) Chlorophyll content and stay green: Chlorophyll content and stay green traits have been found to be associated with heat stress tolerance.,. Xu et al. identified three QTLs for chlorophyll content (Chl1, Chl2, and Chl3) (coincided with three stay-green QTL regions (Stg1, Stg2, and Stg3)) were identified in Sorghum. The Stg1 and Stg2 regions also contain the genes for key photosynthetic enzymes, heat shock proteins, and an abscisic acid (ABA) responsive gene.

(f) Photosynthesis: Decliend photosynthesis is suggested as measure of heat stress sensitivity in plants.

(g) Stem reserve remobilization.

Combination Breeding and Physiological Breeding

The physiological-trait-based breeding approach has merit over breeding for yield *per se* because it increases the probability of crosses resulting in additive gene action. The concept of combination phenomics comes from the idea that two or more stress have common physiological effect or common traits - which is an indicator of overall plant health. Similar analogy in human medical terms is high blood pressure or high body temperature or high white blood cells in body is an indicator of health problems and thus we can select healthy people from unhealthy using such a measure. As both abiotic and abiotic stresses can result in similar physiological consequence, tolerant plant can be separated from sensitive plants. Some imaging or infrared measuring techniques can help to speed the process for breeding process. For example, spot blotch intensity and canopy temperature depression can be monitored with canopy temperature depression.

Low Temperature Tolerance in Plants

Crop Plant Resistance to Freezing Stress

Low-temperature tolerance is a complex quantitative character that is expressed following exposure of plants to temperatures that approach freezing. While a large number of studies have been conducted, there is not a general consensus on the mode of gene action controlling the expression of this character. Recessive, additive, partial dominant and overdominant control have all been reported for genes conditioning low-temperature tolerance. The limited resolution of differences in low-temperature tolerance with field and controlled-freeze tests that employ a single minimum stress level have been at least partially responsible for some of the contradictory conclusions reported for genetic studies. For example, consider a hypothetical genetic study where the differences in low-temperature tolerance between two parents are controlled by additive gene action. In this example, if a single minimum temperature of a controlled-freeze test or field trial was -13 °C, most of the F1 and all of the hardy parent population would survive suggesting that there was dominance for low-temperature tolerance. Similarly, a single minimum temperature of -17 °C would suggest that low-temperature tolerance was a recessive character. In this example, use of both -13 °C and -17 °C in a controlled-freeze test could lead to the mistaken conclusion that low-temperature tolerance was determined by dominant gene action under low levels of stress and recessive gene action under high levels of stress. A series of test temperatures that identifies the LT50 (temperature at which 50 percent of the population is killed in a controlled-freeze test) of each population reduces the likelihood of this type of error.

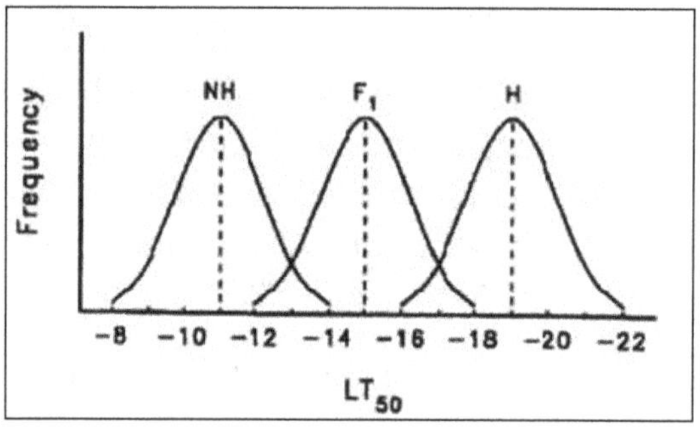

Frequency distributions for parental and F_1 populations from a hypothetical cross where low temperature tolerance is controlled by additive gane action. NH - nonhardy parent (LT_{50}= -11 °C), H-hardy parent (LT_{50}= -19 °C), F_1(LT_{50}= -50 °C).

In spite of the limitations imposed by screening methods, the results of numerous research studies with field crops can be summarized to provide a general picture of the genetics of low-temperature tolerance, the range of genetic variability for gene pools within species, the potential sources of new exploitable genetic variability, and the expression of superior low-temperature tolerance genes introduced into alien genetic backgrounds. The results of these studies demonstrate the complex and multigenic nature of the mechanism controlling low-temperature tolerance.

a) Cytoplasmic factors have been implicated in the control of low-temperature tolerance. However, most studies have concluded that cytoplasmic differences are of minor importance and, if involved at all, play a secondary role in the low-temperature tolerance control mechanism.

b) Genes conferring different levels of low-temperature tolerance are found within and among species. Plant breeders have been able to successfully manipulate this variability to maintain cold hardiness levels of cultivars within established production areas. Considerable variability in the range of low-temperature tolerance also exists among species; however, attempts at interspecific and intergeneric transfers have produced discouraging results.

c) Although there are examples of nonadditive gene action, low-temperature tolerance within species is controlled mainly by genes with additive effects. In wheat, the identification of a dominant gene(s) affecting low-temperature tolerance that is tightly linked to vernalization (Vrn1) and prostrate growth type genes on chromosome 5A has proven to be an important exception to the additive gene action rule. Within the Triticeae, the genes for vernalization are found on the 4th, 5th, and 1st group chromosomes while the 4th and 5th group chromosomes are most commonly associated with low-temperature tolerance. Subsequent work with barley has identified a linkage between low-temperature tolerance and vernalization genes on chromosome 7, which is homeologous with the 5th chromosome group in wheat. These linkages have provided valuable phenotypic markers for the investigation of low-temperature tolerance in Gramineae.

d) Superimposed upon both quality and quantity of low-temperature tolerance genes is the effect of cell size. Smaller cell size amplifies the expression of low-temperature tolerance genes during the acclimation process. In winter wheat, although cold-acclimated cells are smaller, cell size rankings of cultivars follow a similar order for acclimated and nonacclimated plants indicating that differences in this character are intrinsic to the cultivars and not just low-temperature induced. Control of cell size has been localized to the Vrn region of the group 5 chromosomes in wheat.

e) A high degree of genetic balance or harmony is required for full expression of genes central to the low-temperature tolerance mechanism. The unpredictable low-temperature tolerance of artificially synthesized ABD genome hexaploid wheat demonstrates the nonadditivity and asynchronous behaviour of related but unintegrated genetic systems. In synthetic amphiploids of the Triticeae tribe there appears to be a chromosome (gene) dosage effect that favors the expression levels of low-temperature tolerance genes from the parent species contributing the larger chromosome number. Observations made at this level have lead to the suggestion that some degree of genomic integration, which would have been accomplished by recombination and (or) mutation followed by selection, was necessary before maximum low-temperature tolerance was achieved in naturally occurring polyploids of the Triticum - Aegiolops group. These complex interactions should not be unexpected, as there is evidence that certain regulatory mechanisms, such as species specific tRNA and mRNA promoters and interacting transcription factors, may have coevolved in eucaryotic genomes. However, the need for a highly integrated genetic system to maximize gene expression within species does not bode well for efforts to expand gene pools for low-temperature tolerance through interspecific and intergeneric cytogenetic introgressions or the production of transgenics using biotechnological techniques.

f) Specific gene interactions, such as homoeoallelic dominance or threshold effects, may play a role in the final expression of low-temperature tolerance genes introduced into alien genetic backgrounds. As an example, the superior low-temperature tolerance of rye is suppressed when combined in tetraploid and hexaploid wheat backgrounds. These observations once again emphasize the difficulties that can be expected to be associated with efforts to provide breeding programs with superior low-temperature tolerance genes by interspecific and intergeneric transfers using cytogenetic or transformation procedures.

Developmental Regulation

The genetic system that determines low-temperature tolerance can be divided into three separate components. The master switches that are integrated into the mechanism that regulates plant development e.g., vernalization and photoperiod. Low-temperature tolerance genes that have been identified in conventional genetic and cytogenetic studies. Low-temperature induced genes that have been identified through differential screening of cold-acclimated wheat cDNA libraries.

The ability to survive cold winters and continue growth at near freezing temperatures provides a species with a competitive advantage by lengthening the effective growing season or positioning the plant to capitalize on favorable weather periods during the growing season. In order to cope with this stress, plants have evolved reversible acclimation systems that are light and temperature regulated. In areas with long, mild winters, a day length (photoperiod), dormancy, or low-temperature (vernalization) requirement that prevents plants from entering the extremely cold-sensitive reproductive growth stage until the risk of low-temperature damage has passed are the most important adaptive mechanisms. As a result, the evolution and selection of genetic options that permit extensive modification of temperature sensitive metabolic processes and critical structural components is of great concern in the successful adaptation of plants that must survive a wide range of seasonal challenges.

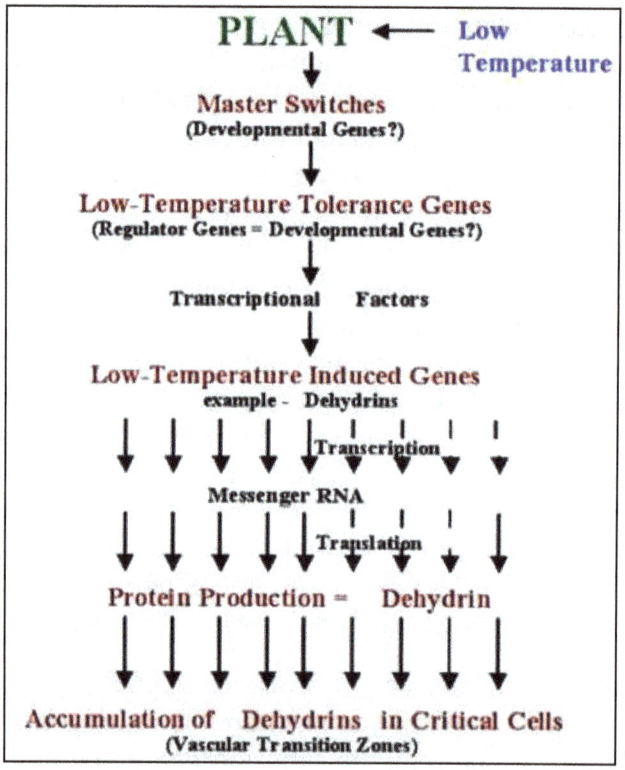

Low-temperature tolerance in cereals is dependent upon a highly integrated system of structural, regulatory, and developmental genes. In regions with cold winters, vernalization requirement is an important adaptive feature that delays heading by postponing the transition from the vegetative to the reproductive phase. Similarly, photoperiod requirement is an adaptation that allows the plant to flower at the optimum time. Time sequence studies have shown that low-temperature gene expression is also developmentally regulated. In this system, transition from the vegetative to the reproductive growth stage is a critical switch that initiates the down regulation of low-temperature induced genes. As a result, full expression of cold hardiness genes only occurs in the vegetative stage and plants in the reproductive phase have a limited ability to cold acclimate. In addition, plants that are still in the vegetative stage have the ability to re-acclimate following periods of exposure to warm temperatures while plants in the reproductive phase have only a limited ability to re-acclimate.

According to the developmental theory, level and duration of gene expression determine the degree of low-temperature tolerance. The developmental genes (vernalization, photoperiod) act as the master switch controlling the duration of expression of low-temperature induced structural genes while the level of low-temperature tolerance is determined by the length of time and degree that the structural genes are up-regulated. Vernalization requirements allow low-temperature genes to be expressed for a longer period of time at temperatures in the acclimation range. Similarly, photoperiod sensitivity allows plants to maintain low-temperature genes in an up-regulated state for a longer period of time under short day compared to long day environments. In both instances, the delay in the transition from the vegetative to the reproductive stage produces increased low-temperature tolerance that is sustained for a longer period of time. This observation also explains why a high level of cold tolerance has not been observed in spring habit cultivars. Because low-temperature gene expression is only up-regulated when the plant is in the vegetative stage, the genetic potentials of spring habit cultivars are not given an opportunity to be fully expressed leaving the impression that the spring habit Vrn1 allele has a dominant pleiotropic effect for frost susceptibility.

Low-temperature Induced Genes

Winter cereals produce several proteins in response to low-temperature stress. Among these low-temperature induced proteins, the dehydrin families have received the most attention in recent low-temperature research (e.g. Wcor 410, Wcs120, dh5, and others). As a group, dehydrins have a wide size range, have no similarity with any enzymes or proteins of known function, are largely hydrophilic, and accumulate to high levels during the late stages of embryogenesis or in response to ABA application, low-temperature, or any environmentally imposed dehydrative force such as drought, intracellular freezing or salinity. For example, the WCOR410 family are peripheral acidic dehydrin proteins found near the plasma membrane and not integral membrane proteins. Immunoelectron microscope analyses of cold acclimated plants has revealed that proteins of this family accumulate in cells of the sensitive vascular transition area where freeze-induced dehydration is likely to be more severe. Their properties, abundance, and localization suggest that they are involved in the protection of critical membranes by replacing water and stabilizing membranes against freezing or dehydration stress.

The low-temperature induced dehydrin gene families have been studied extensively in wheat where they have been mapped to the group 6 chromosomes. Expression studies with the low-temperature induced Wcs120 and Wcor410 dehydrin gene families indicate that, even though there are large differences in low-temperature tolerance, similar proteins are expressed by spring and winter-habit cultivars within species. Cold hardy genotypes just produce more of the same dehydrins than tender genotypes. This indicates a common regulatory control in the level of expression of these structural genes in both hardy and non-hardy genotypes suggesting that the important regulatory factors affecting all low-temperature tolerance associated genes may not be as multigenic in nature as once thought. Further support for this conclusion can be drawn from the report that a single transcriptional factor has been found to regulate the expression of several low-temperature regulated genes in Arabidopsis.

Low-temperature Tolerance Genes/Master Switches

Because of the large number of chromosomes that have been shown to influence low-temperature tolerance in conventional, non-molecular genetic studies, it has been generally assumed that a large number of genes with small effects and complex interactions determine the phenotypic expression of low-temperature tolerance in cereals. However, even with the availability of molecular mapping tools it does not appear that cause and effect relationships will be easily established. To date, molecular mapping studies have only succeeded in locating one low-temperature tolerance gene that appears to be tightly linked to the vernalization gene, vrn1, of chromosome 5A in wheat. Earlier studies had identified a gene in this region with a dominant effect for low-temperature tolerance that was normally expressed in association with the recessive vrn1 allele for winter growth habit in wheat. Vrn1 is homoeoallelic to locus Sh2 in barley and Sp1 in rye, both of which have been linked to genetic differences in low-temperature tolerance.

The group-5 chromosomes carry the vernalization alleles vrn1, vrn4, and vrn3 on chromosomes 5A, 5B, and 5D, respectively. Substitution of each of these chromosomes from a hardy winter cultivar into a nonhardy spring cultivar reduced cell size without affecting growth habit. Genes on the 5th group chromosomes are also known to affect low-temperature tolerance. The Vrn1 region of chromosome 5A and homoeologous loci in wheat and other cereals appear to play an especially important role in determining plant responses to stress. Low-temperature tolerance and other stress tolerance related characters that have been associated with Vrn1 or homoeologous chromosomal regions include: antifreeze protein accumulation, sucrose accumulation, ABA accumulation, unsaturated phospholipid synthesis, prostrate growth habit, cell size, low-temperature tolerance associated with the Wcs120 and Wcor410 gene families, LT50, field survival, flowering time, and fructan content, regulation of drought induced ABA accumulation, and stress tolerance to several minerals. These seemingly complex groupings of tightly linked genes of diverse function can also be explained by the pleiotropic action of regulators that synchronize multi-factorial physical, biochemical and morphological responses of integrated environmentally-induced genetic systems, i.e., master switches.

The function of cells requires that single genes be expressed at different levels and that different genes be expressed at different stages of development and in different tissues. We know this to be intuitively correct because each cell in our body has the genetic information to produce all our body parts yet growth and development proceeds in an orderly, regulated fashion. Similarly, developmental regulation of low-temperature tolerance gene expression provides for a highly integrated system that allows for maximum biochemical and physiological efficiency. An effective system for regulation of low-temperature tolerance responses in plants would also permit multiple use of adaptive mechanisms thereby allowing the successful plant to maintain its competitive advantage by optimizing its metabolic activities. The vrn1 complex and related developmental regulators on chromosome 5A of wheat appear to be an example of this type of highly integrated system.

Role of Transcriptional Factors and Promoters

Genes are expressed through the transcription of DNA sequences to produce mRNAs, which in turn provide the messages that are translated into proteins. Gene expression is driven by a promoter, which is a DNA sequence generally found at the front of the protein coding region of the gene. The promoter determines where, when, and to what extent the gene is expressed. Regulatory elements of various kinds that act as binding sites for unique transcription factors are found throughout these promoters. Regulatory elements can activate or repress transcription, depending on the specific situation. The transcriptional machinery involves several protein factors that interact within the promoter to determine the characteristics of gene expression including where and when the gene is expressed. Promoter sequences found further upstream will also attract specific proteins known as trans-acting or transcription factors, which will activate or repress the transcription machinery in the appropriate cells.

Extrapolation of the present limited knowledge of the genetics of cold hardening in plants suggests that low-temperature tolerance gene expression may be controlled by regulatory elements that act as promoter binding sites for transcription factors that activate the gene. The known low-temperature induced genes carry the same regulatory sequences and they are likely regulated by the same transcriptional factors. The transcription factors vary with environmental conditions and developmental stage and modify gene expression through a combination of transcriptional activation, amplification, repression, and integration that is difficult to predict. For example, activation of the low-temperature induced promoter of cold regulated Arabidopsis genes (cor15a) can be overridden by developmental cues suggesting that developmental genes have an important low-temperature tolerance regulatory role in determining the level of cold-induced gene expression. Involvement of enhancer elements as well as negative and positive regulatory regions for transcriptional regulation in the promoter of low-temperature induced genes have been reported. Multiple DNA binding (repressor) proteins have been associated with promoter regions of plants grown in warm non-acclimating conditions, but these proteins were absent from the promoter in low-temperature acclimated plants indicating a temperature regulated system. These observations point to a system of developmental and pleiotropic regulation that is responsive to temperature cues. They also suggest developmental genes may act as master switches that are capable of regulating the level of expression of cold hardiness genes.

Given the close associations between the vernalization genes and the low-temperature genes, it is possible that vernalization and low-temperature responses are interrelated and the vernalization (vrn) genes located on chromosome 5A may be pleiotropic, regulating both phenological development and the expression of low-temperature tolerance in wheat. Molecular studies designed to investigate these interactions have demonstrated that the regulatory influence exerted by the vrn complex over low-temperature induced structural gene (Wcs120 and Wcor410) expression occurs at the transcriptional level in winter cereals. The wheat group 5 chromosomes, which carry the Vrn genes, have also been found to regulate the expression of a least four low-temperature regulated gene families correlated with low-temperature tolerance. These same chromosomes (particularly 5A) induce higher levels of expression in many low-temperature induced genes that are dispersed across all 3 wheat genomes indicating a single transcriptional activator on chromosome 5A is able to target the low-temperature induced genes. Studies have provided further evidence for regulation of dehydrin genes by group 5 chromosomes. In fact, it has been suggested that barley Sh2 (=Vrn1 homoeoallelic wheat series) may be a dehydrin gene with major regulatory and developmental influence as a result of autoregulatory activity on other dehydrin genes.

The above observations indicate that genes located on chromosome 5A act as a master switch that plays the role of both developmental regulator and transcriptional activator of low-temperature induced genes located on group 6 chromosomes. The master switch theory accommodates the fact that group 5 chromosomes in wheat, and their homoeologous chromosomes in barley and rye, carry both the strongest vernalization genes and the most important low-temperature tolerance genes. It also explains why so many of the stress and low-temperature induced characters appear to be regulated by genes associated with vernalization genes on chromosome 5 and accounts for the often overlooked fact that vernalization and low-temperature acclimation have similar temperature ranges for induction.

Interspecific Gene Transfer and Transformation

While much is known about plant low-temperature response, the maximum cold hardiness potential of most crops has reached a stubborn plateau that has not been breached for decades. In fact, all the efforts of modern science have been unable to produce the super hardy cultivars needed to expand winter crop production into regions requiring a level of cultivar low-temperature tolerance superior to that found in the land races selected by early farmers. The search for superior low-temperature tolerance genes has been expanded to include attempts at interspecific and intergeneric transfers. There are considerable differences in the maximum low-temperature tolerances found in different winter cereals and the possibility that genes can be transferred between species to increase the genetic variability available to winter cereal breeding programs has been explored. However, these attempts have done little more than demonstrate the difficulties that must be overcome before the full potential of superior species-specific cold-tolerance gene expression can be captured through interspecific gene transfers in plant breeding programs.

Rye is the cereal most responsive to low-temperature induction, producing a more rapid rate of cold acclimation and up-regulating low-temperature associated genes to higher levels than other species. Unfortunately, while the structural genes within the Triticeae have a high degree of homology and the regulation of low-temperature tolerance is operational across genomes, we have not been able to successfully exploit the superior low-temperature tolerance genes of rye for cultivar improvement in related cereal species. The superior low-temperature tolerance of rye was suppressed when combined in tetraploid and hexaploid wheat backgrounds. Artificially synthesized ABD genome hexaploid wheat also demonstrated the nonadditivity of closely related genomic systems. Further investigation of low-temperature tolerance gene expression in hybrids among Triticeae species lead to the conclusion that chromosome dosage or ratios influence low-temperature tolerance by shifting competitively balanced systems toward the parent with the greatest chromosome number. Molecular investigations of these hybrids has subsequently revealed that highly conserved and coordinately regulated low-temperature induced gene families of both species are expressed in interspecific crosses. However, these genes were not expressed independently and the degree of low-temperature gene expression in these interspecific crosses was regulated at the transcriptional level by the higher ploidy parent.

These observations indicate that, before we can successfully exploit alien genetic variability for low-temperature tolerance, we must first acquire a greater understanding of the complex genetic mechanisms that plants have evolved for the efficient integration of low-temperature responses into the daily processes of survival, growth, and reproduction. As emphasized earlier, just about every morphological, physiological, and biochemical characteristic that can be measured in the plant changes during low-temperature acclimation. This observation in itself suggests that low-temperature acclimation involves a large number of genes. Threshold induction temperatures, time-temperature relationships for acclimation and deacclimation, effectiveness of regulators, morphological adjustments to changes in light and temperature, and factors that influence the plants transition from the vegetative to the reproductive phase all appear to have an important influence on low-temperature gene regulation in this system. Clearly, low-temperature tolerance gene expression is influenced not only by environment, but also by the pleiotropic effects of other genes. Until biotechnology tools became available it was difficult to separate the genes responsible for low-temperature acclimation and cold tolerance from those associated with metabolic adjustments to low temperature. Molecular genetics has provided researchers with the ability to separate out cause-and-effect relationships and an opportunity to explain some of the apparent contradictions that exist in the low-temperature tolerance literature. This area has been progressing rapidly and we have moved beyond molecular mapping and protein accumulation studies into functional genomics and a more in depth consideration of cause-and-effect relationships with an overall objective of achieving a clearer understanding of the genetic cascade controlling gene expression during acclimation. These investigations have started with a search for the principle mechanisms that winter cereals use to regulate low-temperature gene expression.

Breeding for Resistance to Freezing Stress

A large research effort has been concentrated on solving the mysteries of low-temperature tolerance in plants. Some progress has been made, but the impact of this research on plant breeding has been minimal. While plant breeders have successfully maintained low-temperature tolerance levels of crop plants within established production areas, attempts to produce super-hardy cultivars using the existing genetic variability within species have met with limited success in this century. The primary reasons for this lack of success includes. a) Exploitable genetic variability for low-temperature tolerance has been largely exhausted within the gene pools of most species. b) A large number of genes with small effects and complex interactions are assumed to determine the phenotypic expression of this character making selection difficult. Although most studies indicate that low-temperature tolerance within species is controlled mainly by genes with additive effects; recessive, partial dominant, and overdominant control have also been reported. c) Current methodology for measuring low temperature tolerance gives poor resolution of small phenotypic differences. d) Measures of low-temperature tolerance lack the precision for single plant analysis and many are destructive making selection procedures complicated. e) Poor expression of low-temperature tolerance genes in alien genetic backgrounds has prevented the expansion of gene pools through interspecific and intergeneric transfers e.g., the superior low-temperature tolerance of rye is suppressed in a wheat background.

The ability to acclimate or avoid low-temperature stress varies among species and stages of crop growth. As a consequence, each plant-breeding situation is unique and it is impossible to design a single approach that will satisfy the requirements of all low-temperature tolerance breeding programs. For example, the criteria used to select for low-temperature tolerance are different for actively growing plants that must survive short periods of late spring frost compared to plants that must withstand exposure to long periods of below freezing winter temperatures. When pure, water can supercool or remain unfrozen to its homogeneous nucleation point of approximately -40 °C and, when environmental sources of ice nucleation are absent, supercooling is an important mechanism that determines the level of damage in actively growing plants. Stage of plant development and physical barriers such as the stem nodes, rachis and rachilla that slow ice nucleation and the spread of ice crystals are some of the factors that determine the impact of supercooling on the low-temperature tolerance of tender tissue. Supercooling is also an important survival mechanism for certain cells in woody plants that must survive very low winter temperatures and thermal analysis to detect exotherms resulting from nucleation of supercooled cells have been used to identify differences in the low-temperature tolerance in these species. However, while supercooling is an important low-temperature tolerance mechanism in these instances, winter survival in most plant tissues depends on an ability to accommodate freezing temperatures and prevent damage to the cytoplasm by slowly forming ice crystals in intercellular spaces. Consequently, slow freezing rates and ice nucleation to avoid supercooling are important considerations when conducting freezing tests on most plants that have the ability to low-temperature acclimate. These few examples have been given to emphasize the need for plant breeders to have a thorough understanding of the low-temperature tolerance problems they are addressing. This short introduction also raises a few basic questions that need to be answered to ensure the effectiveness of a low-temperature tolerance-breeding program.

1. What are the climatic conditions of the region that the breeding program targets?
2. What are the critical stress periods for the species under selection?
3. What low-temperature tolerance or avoidance mechanisms are employed and what are the sources of genetic variability in the species under selection?
4. What are the most appropriate methods for identifying superior individuals in breeding populations? Can artificial environments that accurately reproduce the critical stresses experienced in nature be created to facilitate selection programs? Are there low-temperature tolerance prediction tests available that have been field validated for the target region? Do the prediction tests provide measures of field reaction that are accurate enough to be effective as selection tools?
5. Are the necessary resources available to properly implement the selected breeding strategy?

Prediction and Controlled-freeze Tests

The limitations inherent in field survival trials have provided a continuing incentive for researchers to search for rapid, more efficient methods of predicting low-temperature tolerance. Ideally, a low-temperature tolerance screen should be simple, rapid, repeatable, and non-destructive. It should also provide an accurate, precise measure of cold hardiness potential on single plants. The many changes that occur in morphological, biochemical, and physiological characters during low-temperature acclimation have provided fertile ground in the search for prediction tests and differences in several characters, such as plant erectness in winter cereals, tissue water content, cell size, etc., have been shown to be highly correlated with freezing tolerance. Use of these characters as screens satisfies many of the criteria listed above. However, while prediction tests have been used effectively for preliminary low-temperature tolerance selection in breeding programs, high experimental errors limit their usefulness in identifying small differences that are of practical concern to plant breeders, especially where breeding programs target high stress regions.

Controlled-freeze tests on plants that have been cold acclimated in artificial environments are routinely used to measure low-temperature tolerance in physiological and genetic studies. As indicated earlier, rates of freezing and thawing and recovery conditions must reflect the conditions that the plant will experience in nature for controlled-freeze tests to have predictive value. Well designed controlled-freeze tests that employ a single minimum temperature provide a resolution of low-temperature tolerance differences similar to field trials and the same level of caution must be exercised in interpreting results from these two methods. Poor reproduction of cold acclimation rates in repeat experiments makes selection of critical minimum temperatures in controlled-freeze tests almost as difficult as the identification of field sites that yield a high frequency of injury in the target stress range. A series of test temperatures that establish the LT50 (minimum temperature that 50 percent of plants survive) of populations reduces the difficulties associated with identifying critical selection temperatures. Measures of LT50 provide the highest precision and heritability of all low-temperature tolerance prediction tests but, they require a sample of plants from a homogeneous population that can be tested using a series of test temperatures or times. This restricts LT50 measurements to pure lines and limits their usefulness in plant breeding programs that are normally dealing with segregating populations. Increased availability of practical methods for doubled haploid production and asexual propagation, which provide means for quickly producing homogeneous populations, expand the opportunities to use LT50 estimates for selection in plant breeding programs.

The cold tolerance of winter cereal crowns is reduced by prolonged exposure to sub-lethal temperatures and, as a consequence, both temperature and exposure time are important variables in controlled-freeze test procedures. The predicable relationship between time and temperature (Equation 1) has meant that survival time at a constant below freezing temperature can also be used to identify differences in plant low-temperature tolerance.

When combined with the use of controlled environments to simulate conditions for low-temperature acclimation, controlled-freeze-tests provide for greater flexibility in the timing of experiments than field trials. However, while controlled environments should theoretically allow for more rigid control of experimental conditions, comparative studies have shown that field trials usually provide more repeatable results and have lower experimental errors. Consequently, regardless of available resources, field screens are routinely used to provide the final measure of winter survival potential in most breeding programs.

Marker Assisted Selection

The selection for complex genetic traits, such as low-temperature tolerance, can be simplified in plant breeding programs when linked qualitative markers are identified. In wheat, the identification of a dominant gene(s) affecting low-temperature tolerance that is closely associated with the vernalization (vrn1) and prostrate growth habit genes on chromosome 5A has proven to be an important exception to the additive gene action rule. Within the Triticeae, the genes for vernalization are found on the 4th, 5th, and 1st group chromosomes while the 4th and 5th group chromosomes most commonly associated with low-temperature tolerance. Subsequent work with barley has identified a linkage between low-temperature tolerance and vernalization genes on chromosome 7, which is homeologous with the 5th chromosome group in wheat. These linkages have provided valuable phenotypic markers for the investigation of low-temperature tolerance in Gramineae.

The developing science of biotechnology has provided an ever-increasing number of molecular markers that can be used to assist selection in plant breeding programs. However, marker assisted selection requires a detailed linkage map and there is still a dearth of markers for low-temperature tolerance genes. For example, molecular markers in the cereals have been limited to the chromosome regions associated with the homoeoallelic genes Vrn1 in wheat, Sp1 in rye and Sh2 in barley, regions that were previously known to have a significant influence on plant cold hardiness. However, with as many as 15 out of 21 chromosomes in wheat having an influence on low-temperature tolerance, there is still considerable work to be done before the full potential of mapping assisted selection can be realized for this character in most cereal breeding programs.

Advances in biotechnology have provided additional opportunities for plant breeders to expand their attack on the winter-hardiness barrier that has frustrated them for so long. The large number of cold-induced genes associated with low-temperature tolerance and their products that have been isolated in recent years have provided us with a snapshot of these opportunities. For example, antibodies raised against WCS120 proteins produced by low-temperature induced Triticeae genes have provided an immediate practical opportunity to simplify cold hardiness selection procedures in breeding programs. Low-temperature induced expression of this gene family in wheat, as measured by densitometry scanning of Western blots, closely follows changes in LT50 thereby providing a direct means of quantifying phenotypic differences in low-temperature tolerance of cereals.

The rapidly expanding area of functional genomics offers even larger opportunities for the understanding and manipulation of complex genetic systems. As expected in the Triticeae, these investigations have focused on the major low-temperature tolerance genes associated with the vernalization loci. Identification of the main genetic pathways responsible for low-temperature gene expression in one species should provide a guide to charting the routes and establishing the regulation of low-temperature gene expression in other species, especially those that are closely related. Molecular procedures that facilitate the analyses of genome structure and organization for purposes of comparative molecular mapping should allow us to further expand our ability to simultaneously exploit discoveries in a wide range of species. Transformation studies have demonstrated that some of the low-temperature induced genes and their transcriptional factors may be exploited to improve the low-temperature tolerance of tender genotypes. Certainly, we now have the tools to greatly improve our ability to breed for low-temperature tolerance in plants. However, biotechnology has also brought the pharmaceutical model to plant breeding, which includes patents and expanded intellectual property rights that have greatly restricted plant breeders' freedom to operate. These changes have not been unopposed and the full potential for biotechnology to revolutionize the plant breeding world will have to wait for the new order to evolve a more widely accepted and workable plant breeding environment.

Field Selection

In spite of the opportunities offered by cold-hardiness indicators, controlled-freeze testing, and molecular markers, most of today's winter cereal breeding programs still rely heavily on field screening as the final measure of plant winter survival potential. Field-testing is simple, inexpensive, and does not require access to specialized facilities or co-operating programs with conflicting priorities. Unfortunately, the opportunity for selection in field trials only occurs once a year, winters that provide critical selection temperatures usually occur infrequently, and non-uniform stress levels due to variable snow cover and other environmental factors often result in high experimental errors that reduce within-trial selection efficiency.

Where adequate resources are available, a number of measures can be implemented to increase the opportunity for effective field selection for low-temperature tolerance. The frequency of test winters and degree of stress can often be increased by growing trials at or outside the margin of the target production region, thereby increasing the opportunity for selection and the level of selection pressure. Where low-temperature responses have been characterized in detail, computer simulation models and historical weather records may be used to assist in the identification of sites with a high probability of low-temperature stress in the desired range. Methods that control variation in snow cover may be employed to reduce experimental errors and increase the probability of differential injury among breeding lines of plant species with critical meristems near the soil surface. The inclusion of cultivars with known low-temperature tolerance as reference plots to monitor stress levels in trials and the use of a moving average in data analyses provide further opportunities to a) adjust for non-uniform levels of stress, b) maximize the information gleaned from trials with low levels of differential injury, and c) pool results from different trials. The Field Survival Index (FSI) is an example of how this approach has been used to obtain objective measures of low-temperature tolerance in wheat and rye.

Developmental Regulation of Low-temperature Tolerance

Because perennials survive for several years, the staging and regulation of their vegetative/reproductive changes are much more complex than are those of summer and winter annuals. In many cases both vegetative and fully functional reproductive structures must survive the winter, which greatly increases the range of injuries that affect different parts of the plant and the complexity of breeding for low-temperature tolerance in perennials. The biochemical, physiological, and morphological changes associated with low-temperature tolerance clearly interfere with active growth and to be successful a plant must be programmed to recognize and respond to the environmental cues that signal seasonal changes. For these reasons, low-temperature tolerance in perennials normally follows a yearly pattern of environmental cues that include photoperiod and temperature changes. These cues permit the plant to anticipate periods of stress while optimizing its ability to take advantage of favourable periods for growth and reproduction. Consequently, a large part of the plant breeder's task involves selection for the complex genetic mechanisms that plants have evolved for the efficient integration of low-temperature responses into the daily processes of survival, growth, and reproduction.

In general, the longer the plant life cycle the more complex and poorly understood is the low-temperature response mechanism. However, while tree breeders have some of the longest living plants to deal with, they have been much more conscious of the linkage between phenological development and low-temperature response in their selection programs than breeders working on species with shorter life cycles. For example, cereal breeders normally restrict the connection between phenological development and low-temperature tolerance to a consideration of spring and winter growth habit. However, there is strong and growing evidence of close links between the up-regulation of low-temperature tolerance genes and vernalization requirement and photoperiod sensitivity in cereals. For example, cereals with a vernalization requirement or short day photoperiod sensitivity acclimate upon exposure to low-temperature until the point of transition to the reproductive stage after which there is a loss in low-temperature tolerance. Consequently, the point of transition to the reproductive stage is pivotal in the expression of low-temperature tolerance genes. This interaction also makes all low-temperature tolerance associated characters or genes appear to be associated with developmental genes and explains the pleiotropic effects (growth habit and low-temperature tolerance) attributed to developmental genes like vrn1 in wheat.

The linkage of low-temperature tolerance expression to phenological development adapts the plant to the environment for which it was selected or in which it evolved. For example, a high level of low-temperature tolerance is no longer required after the onset of warm conditions in the spring when rapid growth and reproduction begin. Consequently, satisfaction of dormancy, vernalization, and photoperiod requirement results in a decline in low-temperature tolerance of over-wintering plants. In fact, for species adapted to regions with long, mild winters, a high level of freezing tolerance is often less important than a rigorous photoperiod, dormancy, or vernalization requirement that prevents plants from entering the extremely cold-sensitive reproductive growth stage until the risk of low-temperature damage has passed. Consequently, the evolution of, and selection for, genetic options that permit extensive modification of thermosensitive metabolic processes and critical structural components should not come as a surprise, especially in winter annual and perennial plants that must adapt to a wide range of seasonal challenges. It is this genetic system and its regulation and complex interaction with the environment that makes breeding for low-temperature tolerance a continuing challenge.

References

- Methods-used-in-self-pollinated-crops-breeding-methods, plant-breeding: biologydiscussion.com, Retrieved 30 March, 2019
- Rosyara, U. R., S. Subdedi, R. C. Sharma and E. Duveiller. 2010. Photochemical Efficiency and SPAD Value as Indirect Selection Criteria for Combined Selection of Spot Blotch and Terminal Heat Stress in Wheat. Journal of Phytopathology Volume 158, Issue 11-12, pages 813–821, December 2010
- Breeding-methods-used-for-cross-pollinated-crops, vegetable-breeding: biologydiscussion.com, Retrieved 20 July, 2019
- Waghmare, S. G., Sindhumole, P., Shylaja, M. R., Mathew, D., Francies, R. M., Abida, P. S., & Sajini, S. (2018). Analysis of simple sequence repeat (SSR) polymorphism between N22 and Uma rice varieties for marker assisted selection. Electronic Journal of Plant Breeding, 9(2), 511-517

6

Plant Genetic Modification

The direct manipulation of an organism's genes using biotechnology is referred to as genetic modification, or genetic engineering. Genetically modified plants are known as transgenic plants. Plants are genetically modified for various reasons such as for ornamental use, to increase yield in crops and for conservation purposes. This chapter discusses in detail the theories and methodologies related to genetic modification in plants.

Genetic modification is the process of altering the genetic makeup of an organism. This has been done indirectly for thousands of years by controlled, or selective, breeding of plants. Modern biotechnology has made it easier and faster to target a specific gene for more-precise alteration of the organism through genetic engineering. The terms "modified" and "engineered" are often used interchangeably in the context of labeling genetically modified. In the field of biotechnology, GMO stands for genetically modified organism, while in the food industry, the term refers exclusively to food that has been purposefully engineered and not selectively bred organisms.

Techniques other than Genetic Engineering

Simple Selection

The easiest method of plant genetic modification used by our nomadic ancestors and continuing today, is simple selection. That is, a genetically heterogeneous population of plants is inspected, and "superior" individuals—plants with the most desired traits, such as improved palatability and yield—are selected for continued propagation. The others are eaten or discarded. The seeds from the superior plants are sown to produce a new generation of plants, all or most of which will carry and express the desired traits. Over a period of several years, these plants or their seeds are saved and replanted, which increases the population of superior plants and shifts the genetic population so that it is dominated by the superior genotype. This very old method of breeding has been enhanced with modern technology.

An example of modern methods of simple selection is marker-assisted selection, which uses molecular analysis to detect plants likely to express desired features, such as disease resistance to one or more specific pathogens in a population. Successfully applying marker-assisted selection allows a faster, more efficient mechanism for identifying candidate individuals that may have "superior traits."

Superior traits are those considered beneficial to humans, as well as to domesticated animals that consume a plant-based diet; they are not necessarily beneficial to the plant in an ecological or evolutionary context. Often traits considered beneficial to breeders are detrimental to the plant from the standpoint of environmental fitness. For example, the reduction of unpalatable chemicals in a plant makes it more appealing to human consumers but may also attract more feeding by insects and other pests, making it less likely to survive in an unmanaged environment. As a result, cultivated crop varieties rarely establish populations in the wild when they escape from the farm. Conversely, some traits that enhance a plant's resistance to disease may also be harmful to humans.

Crossing

Crossing occurs when a plant breeder takes pollen from one plant and brushes it onto the pistil of a sexually compatible plant, producing a hybrid that carries genes from both parents. When the hybrid progeny reaches flowering maturity, it also may be used as a parent.

Plant breeders usually want to combine the useful features of two plants. For example, they might add a disease-resistance gene from one plant to another that is high-yielding but disease-susceptible, while leaving behind any undesirable genetic traits of the disease-resistant plant, such as poor fertility and seed yield, susceptibility to insects or other diseases, or the production of antinutritional metabolites.

Because of the random nature of recombining genes and traits in crossed plants, breeders usually have to make hundreds or thousands of hybrid progeny to create and identify those few that possess useful features with a minimum of undesirable features. For example, the majority of progeny may show the desired disease resistance, but unwanted genetic features of the disease-resistant parent may also be present in some. Crossing is still the mainstay of modern plant breeding, but many other techniques have been added to the breeders' tool kit.

Interspecies Crossing

Interspecies crossing can take place through various means. Closely related species, such as cultivated oat (Avena sativa) and its weedy relative wild oat (Avena fatua), may cross-pollinate for exchange of genetic information, although this is not generally the case. Genes from one species also can naturally integrate into the genomes of more distant relatives under certain conditions. Some food plants can carry genes that originate in different species, transferred both by nature and by human intervention. For example, common wheat varieties carry genes from rye. A common potato, Solanum tuberosum, can cross with relatives of other species, such as S. acaule.

Chromosome engineering is the term given to nonrecombinant deoxyribonucleic acid (rDNA) cytogenetic manipulations, in which portions of chromosomes from near or distant species are recombined through a natural process called chromosomal translocation. Sears pioneered the human exploitation of this process, which proved valuable for transferring traits that were otherwise unattainable, such as pest or disease resistance, into crop species. However, because transferring large segments of chromosomes also transferred a number of neutral or detrimental genes, the utility of this technique was limited.

Recent refinements allow plant breeders to restrict the transferred genetic material, focusing more on the gene of interest. As a result, chromosome engineering is becoming more competitive with rDNA technology in its ability to transfer relatively small pieces of DNA. Several crop species, such as corn, soybean, rice, barley, and potato, have been improved using chromosome engineering.

Embryo Rescue

Sometimes human technical intervention is required to complete an interspecies gene transfer. Some plants will cross-pollinate and the resulting fertilized hybrid embryo develops but is unable to mature and sprout. Modern plant breeders work around this problem by pollinating naturally and then removing the plant embryo before it stops growing, placing it in a tissue-culture environment where it can complete its development. Such embryo rescue is not considered genetic engineering, and it is not commonly used to derive new varieties directly, but it is used instead as an intermediary step in transferring genes from distant, sexually incompatible relatives through intermediate, partially compatible relatives of both the donor and recipient species.

Somatic Hybridization

Recent advances in tissue-culture technologies have provided new opportunities for recombining genes from different plant sources. In somatic hybridization, a process also known as cell fusion, cells growing in a culture medium are stripped of their protective walls, usually using pectinase, cellulase, and hemicellulase enzymes. These stripped cells, called protoplasts, are pooled from different sources and, through the use of varied techniques such as electrical shock, are fused with one another.

When two protoplasts fuse, the resulting somatic hybrid contains the genetic material from both plant sources. This method overcomes physical barriers to pollen-based hybridization, but not basic chromosomal incompatibilities. If the somatic hybrid is compatible and healthy, it may grow a new cell wall, begin mitotic divisions, and ultimately grow into a hybrid plant that carries genetic features of both parents. While protoplast fusions are easily accomplished, as almost all plants (and animals) have cells suitable for this process, relatively few are capable of regenerating a whole organism, and fewer still are capable of sexual reproduction. This non-genetic engineering technique is not common in plant breeding as the resulting range of successful, fertile hybrids has not extended much beyond what is possible using other conventional technologies.

Somaclonal Variation

Somaclonal variation is the name given to spontaneous mutations that occur when plant cells are grown in vitro. For many years plants regenerated from tis-sue culture sometimes had novel features. It was not until the 1980s that two Australian scientists thought this phenomenon might provide a new source of genetic variability, and that some of the variant plants might carry attributes of value to plant breeders.

Through the 1980s plant breeders around the world grew plants in vitro and scored regenerants for potentially valuable variants in a range of different crops. New varieties of several crops, such as flax, were developed and commercially released. Molecular analyses of these new varieties were not required by regulators at that time, nor were they conducted by developers to ascertain the nature of the underlying genetic changes driving the variant features. Somaclonal variation is still used by some breeders, particularly in developing countries, but this non-genetic engineering technique has largely been supplanted by more predictable genetic engineering technologies.

Mutation Breeding: Induced Chemical and X-ray Mutagenesis

Mutation breeding involves exposing plants or seeds to mutagenic agents (e.g., ionizing radiation) or chemical mutagens (e.g., ethyl methanesulfonate) to induce random changes in the DNA sequence. The breeder can adjust the dose of the mutagen so that it is enough to result in some mutations, but not enough to be lethal. Typically a large number of plants or seeds are mutagenized, grown to reproductive maturity, and progeny are derived. The progeny are assessed for phenotypic expression of potentially valuable new traits.

As with somaclonal variation, the vast majority of mutations resulting from this technique are deleterious, and only chance determines if any genetic changes useful to humans will appear. Other than through varying the dosage, there is no means to control the effects of the mutagen or to target particular genes or traits. The mutagenic effects appear to be random throughout the genome and, even if a useful mutation occurs in a particular plant, deleterious mutations also will likely occur. Once a useful mutation is identified, breeders work to reduce the deleterious mutations or other undesirable features of the mutated plant. Nevertheless, crops derived from mutation breeding still are likely to carry DNA alterations beyond the specific mutation that provided the superior trait.

Induced-mutation crops in most countries (including the United States) are not regulated for food or environmental safety, and breeders generally do not conduct molecular genetic analyses on such crops to characterize the mutations or determine their extent. Consequently, it is almost certain that mutations other than those resulting in identified useful traits also occur and may not be obvious, remaining uncharacterized with unknown effects.

Worldwide, more than 2,300 different crop varieties have been developed using induced mutagenesis, and about half of these have been developed during the past 15 years. In the United States, crop varieties ranging from wheat to grapefruit have been mutated since the technique was first used in the 1920s. There are no records of the molecular characterizations of these mutant crops and, in most cases, no records to retrace their subsequent use.

Cell Selection

Several commercial crop varieties have been developed using cell selection, including varieties of soybeans canola, and flax. This process involves isolating a population of cells from a so-called "elite plant" with superior agricultural characteristics. The cells are then excised and grown in culture. Initially the population is genetically homogeneous, but changes can occur spontaneously (as in somaclonal variation) or be induced using mutagenic agents. Cells with a desired phenotypic variation may be selected and regenerated into a whole plant. For example, adding a suitable amount of the appropriate herbicide to the culture medium may identify cells expressing a novel variant phenotype of herbicide resistance. In theory, all of the normal, susceptible cells will succumb to the herbicide, but a newly resistant cell will survive and perhaps even continue to grow. An herbicide-resistant cell and its derived progeny cell line thus can be selected and regenerated into a whole plant, which is then tested to ensure that the phenotypic trait is stable and results from a heritable genetic alteration. In practice, many factors influence the success of the selection procedure, and the desired trait must have a biochemical basis that lends itself to selection in vitro and at a cellular level.

Breeders cannot select for increased yield in cell cultures because the cellular mechanism for this trait is not known. The advantage of cell selection over conventional breeding is the ability to inexpensively screen large numbers of cells in a petri dish in a short time instead of breeding a similar number of plants in an expensive, large field trial conducted over an entire growing season.

Like somaclonal variation, cell selection has largely been superceded by recombinant technologies because of their greater precision, higher rates of success, and fewer undocumented mutations.

Genetic Engineering

Genetic engineering is a type of genetic modification that involves an intended targeted change in a plant or animal gene sequence to effect a specific result through the use of rDNA technology. A variety of genetic engineering techniques are described in the following text.

Microbial Vectors

Agrobacterium tumefaciens is a naturally occurring soil microbe best known for causing crown gall disease on susceptible plant species. It is an unusual pathogen because when it infects a host, it transfers a portion of its own DNA into the plant cell. The transferred DNA is stably integrated into the plant DNA, and the plant then reads and expresses the transferred genes as if they were its own. The transferred genes direct the production of several substances that mediate the development of a crown gall.

Among these substances is one or more unusual nonprotein amino acids, called opines. Opines are translocated throughout the plant, so food developed from crown gall-infected plants will carry these opines. In the early 1980s strains of Agrobacterium were developed that lacked the disease-causing genes but maintained the ability to attach to susceptible plant cells and transfer DNA.

By substituting the DNA of interest for the crown gall disease-causing DNA, scientists derived new strains of Agrobacterium that deliver and stably integrate specific new genetic material into the cells of target plant species. If the transformed cell then is regenerated into a whole fertile plant, all cells in the progeny also carry and may express the inserted genes. Agrobacterium is a naturally occurring genetic engineering agent and is responsible for the majority of GE plants in commercial production.

Microprojectile Bombardment

Klein and colleagues (1987) discovered that naked DNA could be delivered to plant cells by "shooting" them with microscopic pellets to which DNA had been adhered. This is a crude but effective physical method of DNA delivery, especially in species such as corn, rice, and other cereal grains, which Agrobacterium does not naturally transform. Many GE plants in commercial production were initially transformed using microprojectile delivery.

Electroporation

In electroporation, plant protoplasts take up macromolecules from their surrounding fluid, facilitated by an electrical impulse. Cells growing in a culture medium are stripped of their protective walls, resulting in protoplasts. Supplying known DNA to the protoplast culture medium and then applying the electrical pulse temporarily destabilizes the cell membrane, allowing the DNA to enter the cell. Transformed cells can then regenerate their cell walls and grow to whole, fertile transgenic plants. Electroporation is limited by the poor efficiency of most plant species to regenerate from protoplasts.

Microinjection

DNA can be injected directly into anchored cells. Some proportion of these cells will survive and integrate the injected DNA. However, the process is labor intensive and inefficient compared with other methods.

Transposons/Transposable Elements

The genes of most plant and some animal (e.g., insects and fish) species carry transposons, which are short, naturally occurring pieces of DNA with the ability to move from one location to another in the genome. Barbara McClintock first described such transposable elements in corn plants during the 1950s. Transposons have been investigated extensively in research laboratories, especially to study mutagenesis and the mechanics of DNA recombination. However, they have not yet been harnessed to deliver novel genetic information to improve commercial crops.

Nontransgenic Molecular Methods of Manipulation

Genetic features can be added to plants and animals without inserting them into the recipient organism's native genome. DNA of interest may be delivered to a plant cell, expressing a new protein—and thereby a new trait—without becoming integrated into the host-cell DNA. For example, virus strains may be modified to carry genetic material into a plant cell, replicate, and thrive without integrating into the host genome. Without integration, however, new genetic material may be lost during meiosis, so that seed progeny may not carry or express the new trait.

Many food plants are perennials or are propagated by vegetative means, such as grafting or from cuttings. In these cases the virus and new genes would be maintained in subsequent, nonsexually generated populations. Technically such plants are not products of rDNA because there is no recombination or insertion of introduced DNA into the host genome. Although these plants are not GE, they do carry new DNA and new traits. No such products are known to be currently on the market in the United States or elsewhere.

Transgenic Plants

Transgenic plants or genetically modified plants are plants whose DNA is modified using genetic engineering techniques. In most cases the aim is to introduce a new trait to the plant which does not occur naturally in this species. Examples include resistance to certain pests, diseases or environmental conditions, or the production of a certain nutrient or pharmaceutical agent.

A transgenic crop plant contains a gene or genes which have been artificially inserted instead of the plant acquiring them through pollination. The inserted gene sequence, known as the transgene, may come from another unrelated plant or from a completely different species. Plants containing transgenes are often called genetically modified or GM crops.

Reasons for Making Transgenic Crop Plants

i. The process of transgenic plant development primarily aims at assembling a combination of genes in a crop plant which will make it as useful and productive as possible. Depending on where and for what purpose the plant is grown, desirable genes may provide features such as higher yield or improved quality, pest or disease resistance, or tolerance to heat, cold and drought.

ii. Combining the best genes in one plant is a long and difficult process, especially as traditional plant breeding has been limited to artificially crossing plants within the same species or with closely related species to bring different genes together.

For example, a gene for protein in soybean could not be transferred to a completely different crop such as corn using traditional techniques. Transgenic technology enables plant breeders to bring together in one plant useful genes from a wide range of living sources, not just from within the crop species or from closely related plants.

iii. This technology provides the means for identifying and isolating genes controlling specific characteristics in one kind of organism, and for moving copies of those genes into another quite different organism, which will then also have those characteristics.

iv. This powerful tool enables plant breeders to do what they have always done-to generate more useful and productive crop varieties containing new combinations of genes-but it expands the possibilities beyond the limitations imposed by traditional cross-pollination and selection techniques.

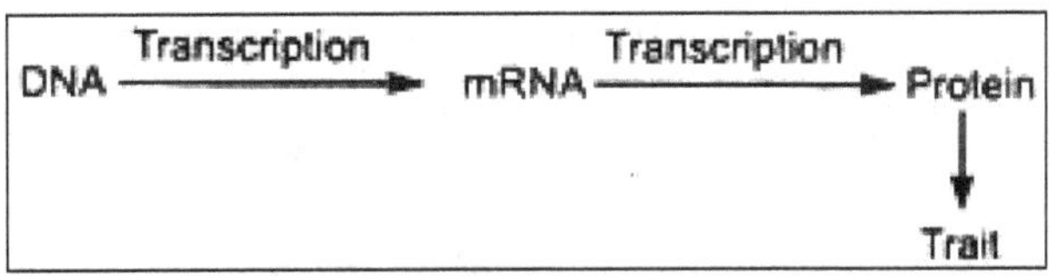

The Universal concept of expression of biological traits

Fundamentals of Transgenic Plant Development

The underlying reason that transgenic plants can be constructed is the universal presence of DNA (deoxyribonucleic acid) in the cells of all living organisms. This molecule stores the organism's genetic information and orchestrates the metabolic processes of life. Genes are discrete segments of DNA that encode the information necessary for assembly of a specific protein.

A specific protein (or an enzyme) encodes for a particular trait. In the production of a transgenic plant our primary aim is to transfer a foreign gene, encoding for some novel traits, into the genome of the plant stably.

After transferring we also need the transgene to integrate and express in the plant's cells. This process as a whole generates a new variety of plant which is new in its own kind and interests us in its massive large scale cultivation.

Four Steps involved in the Production of Transgenic Plants

The fundamental steps involved in the transgenic plant production are as follows:

Step 1: Identifying, Isolation and Cloning of Genes for Agriculturally Important Traits

The very first step in the generation of transgenic plant is to identify and isolate the novel transgene that we want to transfer into the genome of the target plant. Usually, identifying a single gene involved with a trait is not sufficient. We also have to understand how the gene is regulated, what other effects it might have on the plant, and how it interacts with other genes active in the same biochemical pathway.

Step 2: Designing Gene Construct for Insertion

After entering the plant cell the transgene must inter-grate into the genome of the plant stably and express itself successfully so as to produce higher amount of transgenic protein which will be indirectly reflected in the trait controlled by it.

achieve this we have to design a "gene construct" or gene-set, having all the DNA segments necessary to achieve the integration and expression of the transgene. Once a gene has been isolated and cloned (amplified in a bacterial vector), it must undergo several modifications before it can be effectively inserted into a plant.

A gene-set which will be transferred to the target plant has following segments:

A Promoter Sequence: This must be added for the gene to be correctly expressed (i.e., translated into a protein product). The promoter is the on/off switch that controls when and where in the plant the gene will be expressed. To date, most promoters in transgenic crop varieties have been "constitutive", i.e., causing gene expression throughout the life cycle of the plant in most tissues.

The most commonly used constitutive promoter is CaMV 35S, from the cauliflower mosaic virus, which generally results in a high degree of expression in plants. Other promoters are more specific and respond to cues in the plant's internal or external environment. An example of a light-inducible promoteris the promoter from the cab gene, encoding the major chlorophyll a/b binding protein.

The Transgene: Sometimes, the transgene is modified to achieve greater expression in a plant. For example, the Bt gene for insect resistance is of bacterial origin and has a higher percentage of A-T nucleotide pairs compared to plants, which prefer G-C nucleotide pairs. In a clever modification, researchers substituted A-T nucleotides with G-C nucleotides in the Bt gene without significantly changing the amino acid sequence. The result was enhanced production of the gene product in plant cells.

Termination Sequence: This signals to the cellular machinery that the end of the gene sequence has been reached.

A Selectable Marker Gene: This is added in order to identify plant cells or tissues that have successfully integrated the transgene. This is necessary because achieving incorporation and expression of transgenes in plant cells is a rare event, occurring in just a few percent of the targeted tissues or cells.

Selectable marker genes encode proteins that provide resistance to agents that are normally toxic to plants, such as antibiotics or herbicides. Only plant cells that have integrated the selectable marker gene will survive when grown on a medium containing the appropriate antibiotic or herbicide.

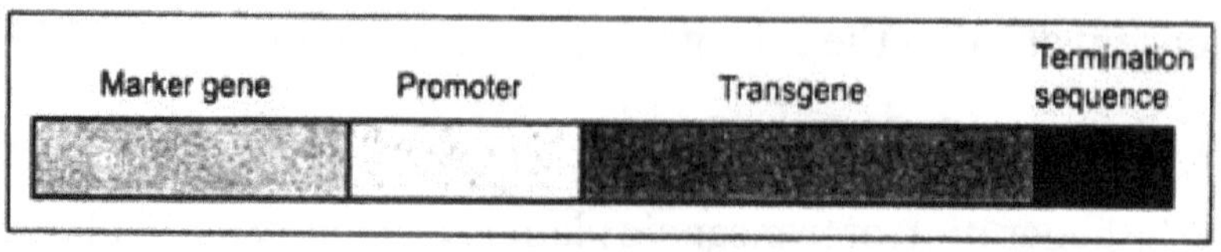

The gone construct that has to be transferred into the target paint.

The essential features of an ideal reporter gene are:

- Lack of endogenous activity in plant cells of the concerned enzyme,
- An efficient and easy detection, and
- A relatively rapid degradation of the enzyme.

The commonly used selectable marker genes include those conferring resistance to the antibiotics kanamycin (nptII, encoding neomycin phosphotransferase) and hygromycin (hptIV, encoding hygromycin phosphotransferase, isolated from E. coli); and broad range herbicides glyphosate (modified versions of the enzyme EPSPS, 5-enolpyruvate shikimate-3-phosphate synthase, isolated from E. coli or Salmonella typhimurium), phosphinothricin (bar, isolated from Streptomyces hygroscopicus, codes for phophinothricin acetyltransferase), etc.

Step 3: Transforming Target Plants with the Gene Construct

There are two Ways of Genetically Transforming the Target Plant:

- Vector mediated gene transfer, and
- Vector less or direct gene transfer.

Step 4: Selection of The Transgenic Plant Tissue/Cells

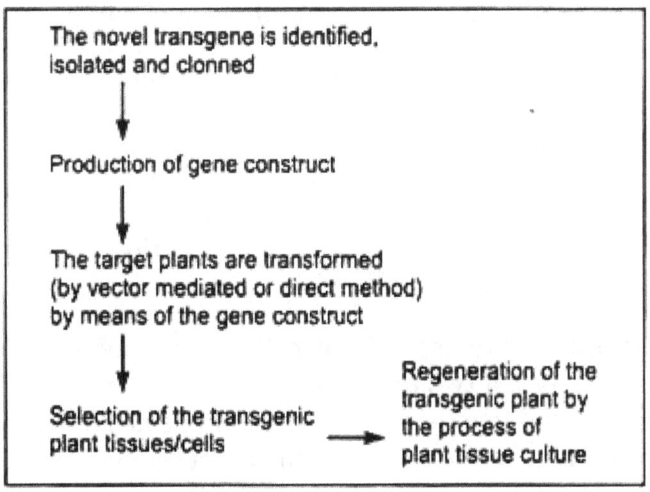

Basic involved in the production of transgenic plants.

Following the gene insertion process, plant tissues are transferred to a selective medium containing an antibiotic or herbicide, depending on which selectable marker was used. Only plants expressing the selectable marker gene will survive and it is assumed that these plants will also possess the transgene of interest.

Regeneration of the Transgenic Plants

To obtain whole plants from transgenic tissues, they are grown under controlled environmental conditions in a series of media containing nutrients and hormones by the process of plant tissue culture.

Table: Promoters used in construction of gene construct.

Promoter	Source	Relative Activity
35S	CaMV 35S RNA gene	Consitutive , high activity : most commonly in dicots
35S + Adh1-1 1	35s promoter + first intron of maize Adh 1 gene	Enhanced promoter activity constitutive
35S + sh1-I 1	35s promoters + first intron of maiz shrunken 1 gene	Better then 35s + adhI -1I in mono – cots constitutive
Adh1	Promoter of alcohol dehydrogenese gene of maiz	Moderate activity in cereals anaerobic expression
Emu	Modified from adh1 promoter and its first intron	Moderate activity in cereals anaerobic expression
Act1 + Actd 1	Rice actin gene + its first intron	Moderate activity constitutive
ubI1+ ubi1 - I 1 (or I 6)	Maize ubiquitin 1 gene promoter + its first (or sixth) intron	high activity in cereals . constitutive
Vicilin promoter	Pea viclin storage protein gene	Seed specific promoters

Integration of the Transgene in the Genome of the Target Plant

In general, transgenes integrate at random sites in any of the chromosomes of the genome of host cells. Usually, in a given cell, integration occurs at a single location. As a result, different cells may be expected to show integration of the transgene at different chromosomal locations.

The number of copies integrated per genome ranges from one to several hundred. In general, multiple copies are integrated when large amounts of DNA are used for transfection, while single copies are integrated with smaller amounts.

When multiple copies are integrated, they are mostly integrated at one site joined to each other head-to-tail, i.e., as a concatemer. However, in a small proportion of cases, the multiple copies are located at several sites in the same genome.

The mechanism of random integration is not known. The entire gene construct, including the vector DNA, becomes integrated. When two different gene constructs are mixed and used for transfection, they tend to be integrated together at the same site; this is known as co-transfection. The sequences flanking a gene on either side influence the expression of this gene.

Therefore, the same transgene integrated at different locations in the genome may show different levels of expression; this is known as position effect. Transgene integration frequently leads to various forms of rearrangements, e.g., duplication, deletion, etc., near the site of integration.

If these changes are large enough, the host gene located at the site of integration may become non-functional. A host gene would also become non-functional if the transgene becomes integrated within the coding region of this gene. When integration of a transgene leads to the loss of function of a host gene, it is called insertional mutagenesis; it often produces aberrant phenotypes.

Analysis of Transgene Integration

The integration of transgene into the genome is confirmed by Southern hybridization of genomic

DNA extracted from the considered transgenic individuals. The DNA is digested with a suitable restriction enzyme prior to electrophoresis.

By choosing appropriate restriction enzymes for DNA digestion, not only the integration of transgene can be established beyond doubt, but information on the number of copies per cell, the orientations of tandemly arranged copies and the presence of single or multiple integration sites is also obtained from Southern hybridization. All the individuals that give positive result with Southern hybridization are regarded as confirmed transgenic.

Detection of mRNA Expression

The mRNAs produced by transgenes is most readily detected if they are with unique sequences, which have no counterparts among those produced by the host genome. A high purity RNA preparation is obtained from the appropriate tissue of transgenic individuals, and is subjected to RNA dot blot hybridization with a radioactive probe specific for the transgene.

Alternatively, the RNA preparation may be used for northern hybridization, which provides additional information on transcript size as well.

Inheritance of Transgenes

The transgenes which are stably integrated are inherited in a Mendelian fashion. They are usually dominant. Instability may occur due to point mutation, like methylation, or rearrangements of the T-DNA region. In addition, homologous recombination between copies of the transgene inserted in the same nucleus can also lead to instability of the gene.

Future Development of Transgenic Technology

New techniques for producing transgenic plants will improve the efficiency of the process and will help resolve some of the environmental and health concerns.

Among the expected changes are the following:

a. More efficient transformation, that is, a higher percentage of plant cells will successfully incorporate the transgene.

b. Better marker genes to replace the use of antibiotic resistance genes.

c. Better control of gene expression through more specific promoters, so that the inserted gene will be active only when and where needed.

d. Transfer of multi-gene DNA fragments to modify more complex traits.

Applications of Transgenic Plants

The six applications of transgenic plants are:

1. Resistance to Biotic Stresses

Genetic engineering of plants has led to the development of crops with increased resistance to biotic stresses which is described in three major categories.

Insect (PEST) Resistance

It is estimated that about 15% of the world's crop yield is lost to insects or pests. A selected list of the common insects and the crops damaged is given in table.

Table: A selected list of common Insect pests along with the major crops damaged by them.

Common name of pest	Botanical name	Crop(s) damaged
Cotton bollworm	Helicoverpa zea	Cotton
Cotton leafworm	Spodoptera littiralis	Rice, maize, cotton, tobacco
Tobacco homworm	Manduca sexta	Tobacco, tomato, potato
European corn borer	Ostrinia nubilalis	Maize
Locust	Locusta migratoria	Grasses
Tobacco budworm	Heliothis virescens	Tobacco, cotton
Tomato fruitworm	Heliothis armigera	Tomato, cotton
Cowpea seed beetle	Callosobruchus maculatus	Cowpea, soybean
Colorado beetle	Leptinotarsa decemlineata	Potato
Brown plant hopper	Nilaparvata lugens	Rice

The damage to crops is mainly caused by insect larvae and to some extent adult insects.

The majority of the insects that damage crops belong to the following orders (with examples):

- Lepidoptera (bollworms.
- Coleoptera (beetles)
- Orthoptera (grasshoppers).
- Homoptera (aphids).

Till some time ago, chemical pesticides are the only means of pest control. Scientists have been looking for alternate methods of pest control for the following reasons (i.e. limitations of pesticide use):

- About 95% of the pesticide sprayed is washed away from the plant surface and accumulates in the soil.
- It is difficult to deliver pesticides to vulnerable parts of plants such as roots, stems and fruits.
- Chemical pesticides are not efficiently degraded in the soil, causing environmental pollution.
- Pesticides, in general, are toxic to non- target organisms, particularly humans and animals.

It is fortunate that scientists have been able to discover new biotechnological alternatives to chemical pesticides thereby providing insect resistance to crop plants. Transgenic plants with insect resistance transgenes have been developed. About 40 genes obtained from microorganisms of higher plants and animals have been used to provide insect resistance in crop plants. Some of the approaches for the bio-control of insects are briefly described.

Resistance Genes from Microorganisms

Bacillus Thuringiensis (Bt) Toxin

Bacillus thuringiensis was first discovered by Ishiwaki in 1901, although its commercial importance was ignored until 1951. B. thuringiensis is a Gram negative, soil bacterium. This bacterium produces a parasporal crystalline proteinous toxin with insecticidal activity. The protein produced by B .thuringiensis is referred to as insecticidal crystalline protein (ICP). ICPs are among the endotoxins produced by sporulating bacteria, and were originally classified as δ-endotoxins (to distinguish them from other classes of α-, β- and γ-endotoxins).

Bt Toxin Genes

Several strains of B. thuringiensis producing a wide range of crystal (cry) proteins have been identified. Further, the structure of cry genes and their corresponding toxin (δ-endotoxin) products have been characterized. The cry genes are classified into a large number of distinct families (about 40) designated as cry 1...... cry 40, based on their size and sequence similarities. And within each family, there may be sub-families. Thus, the total number of genes producing Bt toxins (Cry proteins) is more than 100.

There are differences in the structure of different Cry proteins, besides certain sequence similarities. The molecular weights of Cry proteins may be either large (~130 KDa) or small (~70KDa). Despite the differences in the Cry proteins, they share a common active core of three domains.

Mode of Action of Cry Proteins

Most of the Bt toxins (Cry proteins) are active against Lepidopteran larvae, while some of them are specific against Dipteran and Coleopteran insects. The pro-toxin of Cry I toxin group has a molecular mass of 130 kilo Daltons (130 KDa).

When this parasporal crystal is ingested by the target insect, the pro-toxin gets activated within its gut by a combination of alkaline pH (7.5 to 8.5) and proteolytic enzymes. This results in the conversion of pro-toxin into an active toxin with a molecular weight of 68 KDa.

The active form of toxin protein gets itself inserted into the membrane of the gut epithelial cells of the insect. This result in the formation of ion channels through which there occurs an excessive loss of cellular ATP. As a consequence, cellular metabolism ceases, insect stops feeding, and becomes dehydrated and finally dies.

Some workers i n the recent years suggest that the Bt toxin opens cation-selective pores in the membranes, leading to the inflow of cations into the cells that causes osmotic lysis and destruction of epithelial cells (and finally the death of insect larvae). The Bt toxin is not toxic to humans and animals since the conversion of pro-toxin to toxin requires alkaline pH and specific proteases (These are absent humans and animals).

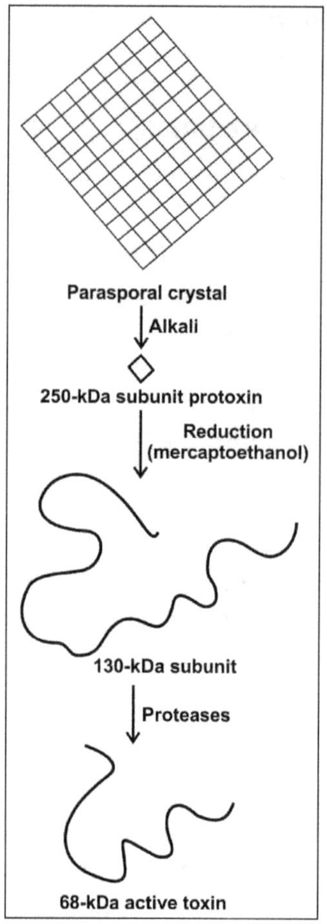

A diagrammatic representation of the formation of active toxin from the parasporal crystal of Bacillus thunngiensis.

Bt Toxin as Bio-pesticide

Preparations of Bt spores or isolated crystals have been used as organic bio-pesticide for about 50 years.

This approach has not met with much Success for the following reasons:

- Low persistence and stability (sunlight degrades toxin) of the toxin on the surface of plants.
- The Bt toxin cannot effectively penetrate into various parts of plants, particularly roots.
- Cost of production is high.

Bt-based Genetic Transformation of Plants

It has been possible to genetically modify (CM) plants by inserting Bt genes and provide pest resistance to these transformed plants. For an effective pest resistance, the bacterial gene in transgenic plants must possess high level expression. This obviously means that the transgene transcription should be under the effective control of promoter and terminator sequences. The early attempts to express cry 1A and cry 3A proteins under the control of CaMV 35S or Agrobacterium T-DNA promoters resulted in a very low expression in tobacco, tomato and potato plants.

Modification of Bt Cry 1A Gene

The wild type transgene Bt cry 1A(b) was found to express at a very low levels in transgenic plants. The nucleotide sequence of this gene was modified (G + C content altered, several polyadenylation signals removed, ATTTA sequence deleted etc). With appropriate sequence changes, an enormous increase (about 100 fold) in the Bt toxin product formation was observed.

The transgenic Bt crops that were found to provide effective protection against insect damage were given approval for commercial planting by USA in the mid-1990s. Some biotechnological companies with their own trade names introduced several transgenic crops into the fields. Among these, only maize and cotton Bt crops are currently in use in USA. The other genetically modified plants met with failure for various reasons.

Table: Bt—based genetically modified crop plants developed for commercial use.

Crop	Trade name	Bt protein	Resistance to insect(s)
Cotton	Bollgard	Cry 1Ac	Cotton bollworm, tobacco budworm
Maize	Yieldgard knockout	Cry 1Ab	European corn borer
Maize	Starlink	Cry 9C	European corn borer
Maize	Herculex I	Cry 1F	European corn borer
Maize	Bt-Xtra	Cry 1Ac	European corn borer
Potato	New-leaf	Cry 3A	Colorado beetle

Advantages of Transgenic Plants with Bt Genes

- Bt genes could be expressed in all parts of the plants, including the roots and internal regions of stems and fruits. This is not possible by any chemical pesticide.
- Toxic proteins are produced within the plants; hence they are environmental-friendly.
- Bt toxins are rapidly degraded in the environment.

Problem of Insect Resistance to Bt crops

The major limitation of Bf-gene possessing transgenic plants is the development of Bt-resistant insects. The Bf toxin is a protein, and the membrane receptor (of the gut) through which the toxin mediates its action is also a protein. It is possible that the appropriate mutations in the insect gene coding for receptor protein may reduce the toxin binding and render it ineffective. This may happen within a few generations by repeated growing of Bt crops. Several approaches are made to avoid the development of resistance in insects:

- Introduction of two different Bf toxin genes for the same target insect.
- Development of transgenic plants with two types of insect resistance genes e.g. Bt gene and proteinase inhibitor gene.
- Rotating Bt crops with non-Bf crops may also prevent the build-up of resistance in insect population.

Environmental Impact of Bt Crops

The most serious impact of Bt crops on environment is the build-up of resistance in the pest population. In 1999, another issue was brought to light about Bt crops. It was reported that the pollen from Bt maize might be toxic to the larvae of Monarch butterfly. This generated considerable opposition to Bt crops by the public, since Monarch butterfly is one of the most colorful natives in USA.

It was later proved that the fears about the impact of Bt crops on the monarch butterfly were without the required scientific evidence. The lesson learnt from the monarch butterfly episode is that the risks of CM crops should be thoroughly assessed before they are reported.

Usage of Bt

The usage Bt is commonly used for a transgenic crop with a cry gene e.g. Bt cotton. In the same way, Cry proteins are also referred to as Bt proteins. It may also be stated here that the authors use four different names for the same group of proteins-δ-endotoxin, insecticidal crystal protein (ICP), Cry and now Bt.

Resistance Genes from other Microorganisms

There are certain other insect resistant genes from other microorganisms. Some of the important ones are listed:

- Cholesterol oxidase of Strepotmyces culture filtrate was found to be toxic to boll weevil larvae. Cholesterol oxidase gene has been introduced into tobacco to develop a transgenic plant.
- Isopentenyl transferase gene from Agrobacterium tumefaciens has been introduced into tobacco and tomato. This gene codes for an important enzyme in the synthesis of cytokinin. The transgenic plants with this transgene were found to reduce the leaf consumption by tobacco hornworm and decrease the survival of peach potato aphid.

Resistance Genes from Higher Plants

Certain genes from higher plants were also found to result in the synthesis of products possessing insecticidal activity. Some authors regard them as non-Bt insecticidal proteins. A selected list of plant insecticidal (non-Bt) genes used for developing transgenic plants with insect resistance is given table. Some of them are briefly described.

Table: A selected list of plant Insecticidal (non-Bt) genes used for developing transgenic plants with insect resistance.

Plant gene	Transgenic plant(s)	Encoded protein	Resistance to incect(s)
Protease inhibitors			
CpTi	Potato, apple, rice, sunflower, wheat, tomato	Trypsin	Coleoptera, Lepidoptera

CII	Tobacco, potato	Serine protease	Coleoptera, Lepidoptera
PF-IV	Potato, tobacco	Serine protease	Lebidoptera
oC-1	Tobacco, oilseed rape	Cysteine protease	Coleoptera, Homoptera
CMe	Tobacco	Trypsin	Lebidoptera
α-Amylase inhibitors			
α-A1-Pv	Pea, tobacco	α-Amylase	Coleoptera
WMAI-1	Tobacco	α-Amylase	Lepidoptera
Lectins			
GNA	Potato, rice, sugarcane, sweet potato, tobacco	Lectin	Homoptera, Lepidoptera
WGA	Maize	Agglutin	Lepidoptera, Coleoptera
Others			
BCH	Potato	Chitinase	Homoptera, Lepidoptera
TDC	Tobacco	Tryptophan decarboxylase	Homoptera

Proteinase (Protease) Inhibitors

Proteinase inhibitors are the proteins that inhibit the activity of proteinase enzymes. Certain plants naturally produce proteinase inhibitors to provide defence against herbivorous insects. This is possible since the inhibitors when ingested by insects interfere with the digestive enzymes of the insect. This results in the nutrient deprivation causing death of the insects. It is possible to control insects by introducing proteinase inhibitor genes into crop plants that normally do not produce these proteins.

Cowpea Trypsin Inhibitor Gene

It was observed that the wild species of cowpea plants growing in Africa were resistant to attack by a wide range of insects. Research findings revealed that insecticidal protein was a trypsin inhibitor that was capable of destroying insects belonging to the orders Lepidoptera (e.g. Heliothis virescans), Orthaptera (e.g. Locusta migratoria) and Coleoptera (e.g. Anthonous grandis).

Cowpea trypsin inhibitor (CpTi) has no affect on mammalian trypsins; hence it is non-toxic to mammals. CpTi gene was introduced into tobacco, potato and oilseed rape for developing transgenic plants. Survival of insects and damage to plants were much lower in plants possessing CpTi gene.

Advantages of Proteinase Inhibitors

- Many insects, not controlled by Bt, can be effectively controlled.
- Use of proteinase gene along with Bt gene will help to overcome Bt resistance development in plants.

Limitations of Proteinase Inhibitors

- Unlike Bt toxin, high levels of proteinase inhibitors are required to kill insects.
- It is necessary that the expression of proteinase inhibitors should be very low in the plant

parts consumed by humans, while the expression should be high in the parts of plants utilized by insects.

α-Amylase Inhibitors

The insect larvae secrete a gut enzyme a- amylase to digest starch. By blocking the activity of this enzyme by α-amylase inhibitor, the larvae can be starved and killed. α-Amylase inhibitor gene (α-AI-Pv) isolated from bean has been successfully transferred and expressed in tobacco. It provides resistance against Coleoptera (e.g. Zabrotes sub-fasciatus).

Lectins

Lectins are plant glycoproteins and they provide resistance to insects by acting as toxins. The lectin gene (CNA) from snowdrop (Calanthus nivalis) has been transferred and expressed in potato and tomato. The major limitations of lectin are that it acts only against piercing and sucking insect, and high doses are required.

Resistance Genes from Animals

Proteinase inhibitor genes from mammals have also been transferred and expressed in plants to provide resistance against insects, although the success in this direction is very limited. Bovine pancreatic trypsin inhibitor (BPTI) and α1– antitrypsin genes appear to be promising to offer insect resistance to transgenic plants.

Insect Resistance through Copy Nature Strategy

Some of the limitations experienced by transferring the insecticidal genes (particularly Bt) and developing transgenic plants have prompted scientists to look for better alternatives. The copy nature strategy was introduced in 1993 (by Boulter) with the objective of insect pest control which is relatively sustainable and environmentally friendly.

The copy nature strategy for the development of insect -resistant transgenic plants has the following stages:

1. Identification of leads: The first step in copy nature strategy is to identify the plants (from world over) that are naturally resistant to insect damage.
2. Isolation and purification of protein: The protein with insecticidal properties (from the resistant plants) is isolated and purified. The sequence of the protein is determined, and the gene responsible for its production identified.
3. Bioassay of isolated protein: The activity of the protein against the target insects is determined by performing a bioassay in the laboratory.
4. Testing for toxicity in mammals: It is absolutely necessary to test the toxicity of the protein against mammals, particularly humans. If the protein is found to have any adverse effect, the copy nature strategy should be discontinued.
5. Gene transfer: By the conventional techniques of genetic engineering, the isolated gene corresponding to the protein toxin is introduced into the crop plants.

6. Selection of transgenic plants: After the gene transfer, the transgenic plants developed should be tested for the inheritance and appropriate expression of transgene. The efficiency of insecticidal protein to destroy insects is also evaluated.

7. Evaluation for biosafety: Field trails have to be conducted to evaluate the crop yield, damage to insects, influence on the environment with respect to the transgenic plant. The copy nature strategy, though time consuming and many a times unsuccessful, takes into account the complex interplay in biological communities (between plants animals, microorganisms) and physical environment.

Virus Resistance

Virus infections of crops may result in retarded cell division (hypoplasia), excessive cell division (hyperplasia), and cell death (necrosis). The overall effects of virus infections are growth retardation, lowered product yield and sometimes complete crop failure. The chemical methods used to control various plant pathogens will be ineffective with respect to plant viruses since the viruses are intracellular obligate parasites.

There are however, certain safe agricultural practices to control/reduce viral infections to plants:

1. Use of seeds that are virus – free.
2. Control insects that spread plant viruses.
3. Control weeds that serve as alternate hosts for viruses.
4. Use cultivars that possess virus resistance.

It is possible to immunize plants against viral damages by expressing viral proteins in the plant cells. With the advances made in genetic engineering, it has become a reality to develop transgenic plants with virus resistance.

This is mostly done by employing virus-encoded genes – virus coat proteins, movement proteins, transmission proteins, satellite RNA, antisense RNAs and ribozymes. In recent years, some attempts are also made to provide virus resistance to plants by using animal genes. Some of the developments are briefly described.

Virus Coat Proteins

The virus coat protein-mediated approach is the most successful one to provide virus resistance to plants. It was in 1986, transgenic tobacco plants expressing tobacco mosaic virus (TMV) coat protein gene were first developed. These plants exhibited high levels of resistance to TMV. Excited by this remarkable success, scientists have worked with many more viruses (around 30 or so) and developed crops with virus coat protein-mediated protection.

A selected list of the virus resistant transgenic plants with sources of virus coat protein genes is given in table. The transgenic plant providing coat protein-mediated resistance to virus are rice, potato, wheat, tobacco, peanut, sugar beet, alfalfa etc. The viruses that have been used include alfalfa mosaic virus (AIMV), cucumber mosaic virus (CMV), potato virus X (PVX), potato virus Y (PVY), citrus tristeza virus (CTV) and R rice stripe virus (RSV).

Table: A selected list of the virus-resistant transgenlc plants with sources of virus protein coat genes.

Plant	*Source(s) of virus coat protein gene*
Tobacco	TMV, CMV, AIMV
Rice	RTSV, RSV, RYMV
Wheat	SBWMV, BYDV
Potato	PVX, PVY, PLRV
Squash	CMV, ZYMV
Sugar beet	BNYVV
Peanut	TSWV
Papaya	PRSV
Citrus	CTV
Alfalfa	AIMV

Advantages of Virus Coat Proteins

The coat protein gene from one virus sometimes provides resistance (cross protection) to some other viruses, which may be unrelated e.g. TMV of tobacco plant provides resistance to potato virus X, alfalfa mosaic virus and cucumber mosaic virus.

Limitation of Virus Coat Proteins

The virus coat protein-mediated protection is successful for viruses with single-stranded RNA genomes. However, this approach is of not much use for viruses with genomes containing double-stranded RNA and single-stranded DNA.

Mechanism of action of Virus Coat Proteins

As the transgenic plant expresses the gene for coat protein of a given virus, the ability of the same virus to infect the plants again is drastically reduced. Despite a remarkable success in the virus coat protein-mediated protection, the molecular mechanism of the protection is not clearly known.

Movement Proteins

As the virus infects the plant cells, its rapid spread through intercellular junctions (plasmadesmata) of vascular tissue occurs through the participation of movement proteins produced by the viruses. Good examples of movement proteins are 30KDa protein of tobacco mosaic virus (TMV) and 32KDa protein of brome mosaic virus (BMV).

Transgenic tobacco plants that express a mutated 30KDa movement protein have been developed. The TMV infection to these plants is much less. It is believed that the mutated movement protein competes with the wild-type TMV-coded protein thereby reducing the spread of the virus (TMV).

In recent years, a recombinant movement protein having the components of golden mosaic virus and African cassava mosaic virus have been developed. This protein effectively interferes with the spread of both the viruses. The advantage with movement protein strategy is that it is applicable to single-stranded DNA viruses (Gemini viruses) also.

Transmission Proteins

There is a good coordination and interaction between plant viruses and insect vectors for the spread of viruses from one plant to another. Certain viral-encoded transmission proteins do this job effectively. It is possible to produce mutated transmission proteins and block the spread of viruses. Thus, the spread of insect-transmitted viruses can be prevented by engineering crops to express a defective virus-transmission protein.

Satellite RNAs

Plant viral satellite RNAs are small RNA molecules that multiply in the host cells with the help of specific helper viruses. These satellite RNAs are encapsulated together with the respective helper viruses. In general, the presence of satellite RNAs reduces the severity of the viral disease and the symptoms, and thus reduces the effect of the virus.

Transgenic plants containing satellite sequences have been developed to provide resistance to virus diseases. One example is given here. When cucumber mosaic virus (CMV) infects pepper plants, severe symptoms appear. These symptoms can be minimized with higher plant yield when CMV is co-inoculated with a satellite RNA.

Satellite RNA approach is not widely used due to several limitations:

- Some of the satellite RNA may increase the severity of disease symptoms in some plants.
- Satellite RNAs mutate very rapidly which may sometimes result in a highly virulent agent.
- Re-combinations between satellite RNAs have been detected. This may lead to serious consequences.

Antisense RNAs

The antisense RNA approach is designed to specifically interfere with virus replication. By use of genetic engineering, a complementary DNA Strand of a gene (DNA sequence) can be inserted in reverse orientation ($3' \rightarrow 5'$ as opposed to $5' \rightarrow 3'$) and this is referred to as antisense gene.

The mRNA produced by antisense gene is complementary to the mRNA synthesized by normal gene. As a result, both these mRNAs hybridize and thus the normal translation of mRNA is blocked. The net effect of employing an antisense gene into a cell is that it blocks a specific gene expression.

It is possible to introduce viral antisense genes into plants and produce mRNAs complementary to viral sequences involved in virus replication. The antisense mRNAs can block the replication of viruses. Initially, antisense RNA approach was carried out in single-stranded RNA viruses. The success of this approach however, was limited probably due to the following reasons:

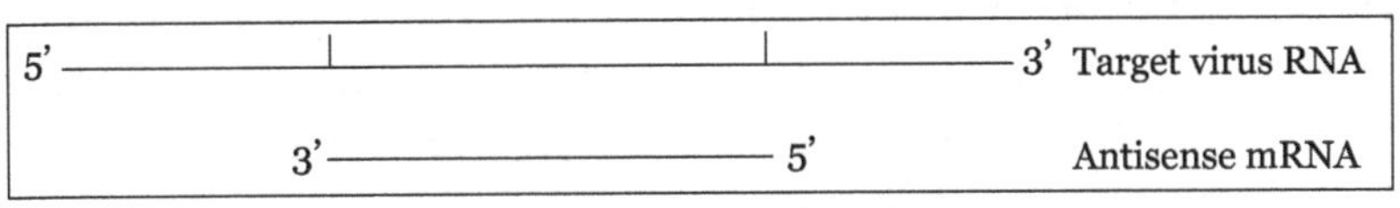

Hybridization of antisense mRana with virus block replication.

- High concentration of antisense mRNA may be required.
- Protein association with mRNA interferes with hybridization (between sense mRNA and antisense mRNA).

Antisense RNA approach may be more useful for DNA viruses. In fact, tomato golden mosaic virus (TGMV) replicase coding sequence was cloned in antisense orientation and introduced into tobacco plants. The transgenic tobacco plants expressed antisense RNA of TGMV replicase. These plants were resistant to TGMV infection.

Ribozymes

Ribozymes are small RNA molecules which promote the catalytic cleavage of RNA. For providing virus resistance, ribozymes in the form of antisense RNAs capable of cleaving the target viral (sense) RNAs have been developed. The ribozyme (antisense RNA) binds to a small sequence of viral RNA and splits.

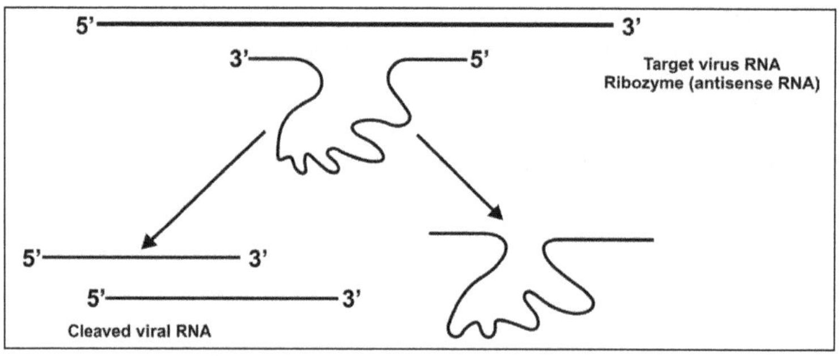

Action of ribozymes as antisense RNAs to block virus replication.

In this way, it is possible to block the replication of viral RNA. However, the ribozymes approach has not been very successful in plants.

Fungal and Bacterial Diseases Resistance

The plants do posses general defence systems against invading pathogens. This is however, not truly comparable with the immune system of the animals. Whenever there is a cellular damage caused by pathogens (fungi, bacteria) and plant pests, the general defence system of plants get geared up to provide some amount of protection to the plant. This natural disease resistance of plants is inadequate.

However, knowledge on the natural systems of plant resistance is useful for the biotechnological approaches to develop disease resistance. Some of the defenses of plants and the biotechnological approaches are briefly described.

Pathogenesis-Related (Pr) Proteins

To defend themselves against the invading pathogens (fungi and bacteria), plants accumulate low molecular weight proteins which are collectively regarded as pathogenesis-related (PR) proteins. The different types of PR proteins and their properties are given in table. Some of the most important ones are described.

Table: Different types of pathogenesis-related (PR) proteins produced in plants.

Type	Properties
PR-1	Antifungal
PR-2	Endo β-1, 3-glucanases
PR-3	Endochitinases
PR-4	Antifungal, endochitinase
PR-5	Antifungal, thaumatin-like proteins, osmotins
PR-6	Protease inhibitors
PR-7	Endoprotease
PR-8	Chitinase/lysozyme
PR-9	Peroxidases
PR-10	Ribonucleases
PR-11	Endochitinase activity
PR-12	Defensins
PR-13	Thionins
PR-14	Lipid transfer proteins (non-specific)

Chitinase

Chitin is a constituent of fungal cell walls which can be hydrolysed by the enzyme chitinase. Certain chitinase genes from plants have been isolated and characterized. A bacterial chitinase gene obtained from a soil bacterium (Serratia marcescens) was introduced and expressed in tobacco leaves. Some other workers isolated a chitinase gene from bean (Phaseolus vulgaris) and developed transgenic plants of tobacco and Brassica napus with this gene. The transformed tobacco plants were found to be resistant to infection of the pathogen Rhizoctonia solani. In case of B. napus, the protection however, was comparatively less.

Glucanase

Glucanase is another enzyme that degrades the cell wall of many fungi. The most widely used glucanase is β-1, 4-glucanase. The gene encoding for β-1, 4-glucanase has been isolated from barley, introduced, and expressed in transgenic tobacco plants. This gene provided good protection against soil-borne fungal pathogen Rhizoctonia solani.

The resistance to fungal pathogens is much higher if both chitinase and glucanase producing genes are present in transgenic plants. By this approach, fungal resistant tobacco, tomato and carrot have been developed.

Ribosome-inactivating Proteins (RIPs)

Ribosome-inactivating proteins offer protection against fungal infections. They act on the large rRNA of eukaryote and prokaryote ribosomes (remove an adenine residue from a specific site), and thus inhibit protein biosynthesis.

Certain RIPs that do not inhibit plant ribosomes were identified and the corresponding genes have been used to develop transgenic plants e.g. Type-I barley RIP is used to provide resistance to fungal infections. Some authors use the term antimicrobial proteins to RIPs. The other examples of antimicrobial proteins are lectins, defensins, lysozyme, thionins etc.

Lysozyme

Lysozyme degrades chitin and peptidoglycan of cell wall and in this way fungal infection can be reduced. Transgenic potato plants with lysozyme gene providing resistance to Eswinia carotovora have been developed.

Defensins

Defensins are antimicrobial peptides (26-50 amino acid residues) found in all the plant cells. They attack the microbial plasma membrane; however this is not adequate to provide resistance to pathogens. In recent years, an artificial defensin gene has been developed and introduced into potatoes. These potatoes developed resistance to the bacterium Eswinia carotovora.

Thionins

Thionin proteins also offer protection against bacteria. Thionin coding genes have been introduced into tobacco and the transgenic plant so developed showed resistance to Pseudomonas syringae.

Phytoalexins

Phytoalexins are secondary metabolites produced in plants in response to infection. They are low-molecular weight and antimicrobial in nature. The phytoalexins usually present in specialized cells or organelles are mobilized when infection occurs. Further, during infection there occurs induction of genes for increased production of phytoalexins.

Stilbene synthase is a key enzyme for the synthesis of a common phytoalexin. The gene coding stilbene synthase has been isolated from peanut and introduced into tobacco, rice and Brassica napus. The transgenic plants carrying stilbene synthase gene were resistant against some fungi. A selected list of transgenic plants developed, along with the genes transferred and the controlled pathogens is given in table.

Table: A selected list of transgenic plants developed along with the genes transferred and the resistance provided against the pathogens (fungus/bacterium).

Crop	Gene(s) transferred	Resistance against pathogen
Pathogenesis-related (PR) proteins		
Tobacco	Chitinase from bacterium (Serratia marcescens) Bean chitinase Chitinase and 1, 3-β glucanase	Alternaria longipes Rhizoctonia solani, Phytophthora parasitica Cercospora nicotinae
Rice	Chitinase	Rhizoctonia solani
Carrot	Chitinase and 1, 3-β glucanase	Alternaria dauci, A. radicina

Tomato	Chitinase and 1, 3-β glucanase	Fusarium oxysporum
Brassica napus	Chitinase	Rhizoctonia solani
Antimicrobial proteins		
Tobacco	Barley ribosome inactivating protein	Rhizoctonia solani
Tobacco	Defensin from radish	Alternaria longipes
Tobacco	α-Thionin gene from barley	Pseudomonas syringae
Potato	Bacteriophage T-4 lysozyme	Erwinia carotovora
Phytoalexins		
Rice	Stilbene synthase	Pyricularia oryzae
Tobacco	Stilbene synthase	Botrytis cinerea

Nematode Resistance

Nematodes are simple worms found in the soil. They possess a complete digestive tract. The annual crop loss of the world due to nematode (roundworm) infestation is very high.Some workers have identified and cloned a nematode resistance gene from wild beet plants.

It is proposed that this gene encodes a protein that detects the pests (nematodes) and triggers a defensive reactions in the plant. It is believed that some chemical compounds that destroy the gut of the nematode are produced. Attempts were also made to transfer the nematode resistance gene to sugar beet. Not much success was reported, the major limitation being the difficulty in cultivating the gene-altered cells of sugar beet.

2. Resistance to Abiotic Stresses

Plants are constantly being subjected to environmental stresses that may result in deterioration of crop plants, and a very low or even no yield. Plants are dependent on the subtle internal mechanisms for tolerance of various stresses.

The in situ tolerance of crop plants, whenever present, is inadequate and therefore cannot give protection against the stresses. A wide range of strategies are required to engineer plants against a particular type of stress tolerance. Some of the abiotic stresses and the recombinant strategies developed to overcome them are described.

Herbicide Resistance

Weeds (wild herbs) are unwanted and useless plants that grow along with the crop plants. Weeds compete with traps for light and nutrients, besides harbouring various pathogens. It is estimated that the world's crop yield is reduced by 10-15% due to the presence of weeds.

To tackle the problem of weeds, modern agriculture has developed a wide range of weed killers which are collectively referred to as herbicides. In general, majority of the herbicides are broad-spectrum as they can kill a wide range of weeds.

A good or an ideal herbicide is expected to possess the following characteristics:

- Capable of killing weeds without affecting crop plants.

- Not toxic to animals and microorganisms.
- Rapidly trans-located within the target plant.
- Rapidly degraded in the soil.

None of the commercially available herbicides fulfills all the above criteria. The major limitation of the herbicides is that they cannot discriminate weeds from crop plants. For this reason, the crops are also affected by herbicides, hence the need to develop herbicide-resistant plants. Thus, these plants provide an opportunity to effectively kill the weeds (by herbicides) without damaging the crop plants.

Strategies for Engineering Herbicide Resistance

A number of biological manipulations particularly involving genetic engineering are in use to develop herbicide-resistant plants.

1. Overexpression of the target protein: The target protein, being acted by the herbicide can be produced in large quantities so that the affect of the herbicide becomes insignificant. Overexpression can be achieved by integrating multiple copies of the genes and/or by using a strong promoter.

2. Improved plant detoxification: The plants do possess natural defense systems against toxic compounds (herbicides). Detoxification involves the conversion of toxic herbicide to non-toxic or less toxic compound. By enhancing the plant detoxification system, the impact of the herbicide can be reduced.

3. Detoxification of herbicide by using a foreign gene: By introducing a foreign gene into the crop plant, the herbicide can be effectively detoxified.

4. Mutation of the target protein: The target protein which is being affected by the herbicide can be suitably modified. The changed protein should be capable of discharging the functions of the native protein but is resistant to inhibition by the herbicide.

Once the resistant target protein gene is identified, it can be introduced into the plant genomes, and thus herbicide-resistant plants can be developed. For success in the development of herbicide resistant plants, good knowledge of the target protein and the action of herbicides is required. Some of the developments made in the herbicide resistance of plant are briefly described.

Glyphosate Resistance

Glyphosate, is a glycine derivative. It acts as a broad-spectrum herbicide and is effective against 76 of the world's worst 78 weeds. Glyphosate is less toxic to animals and is rapidly degraded by microorganisms. In addition, it has a short half-life. The American chemical company Monsanto markets glyphosate as Round up.

Mechanism of Action of Glyphosate

Glyphosate is rapidly transported to the growing points of plants. It is capable of killing the plants even at a low concentration. Glyphosate acts as a competitive inhibitor of the enzyme 5-enoylpyruvylshikimate 3-phosphate synthase (EPSPS). This is a key enzyme in shikimic acid pathway that results in the formation of aromatic amino acids (tryptophan, phenylalanine and tyrosine), phenols and certain secondary metabolites.

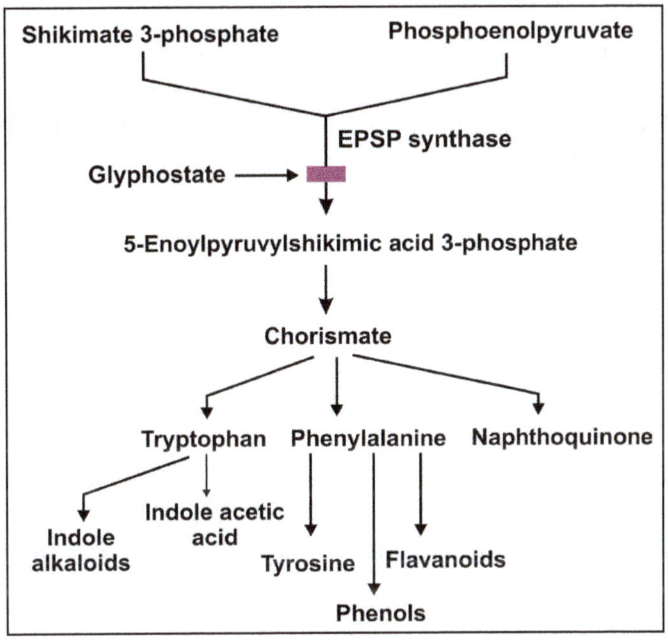

Shikimate pathway indicating the action of the herbicide glyphosate (EPSP synthase-5-enoylpyruvylshikimate 3-phosphate synthase).

The enzyme EPSPS catalyses the synthesis of 5-enoylpyruvylshikimate 3-phosphate from shikimate 3-phosphate and phosphoenoylpyruvate. Glyphosate has some structural similarly with the substrate phosphoenol pyruvate. Consequently, glyphosate binds more tightly with EPSPS and blocks the normal shikimic acid pathway. Thus, the herbicide glyphosate inhibits the biosynthesis of aromatic amino acids and other important products.

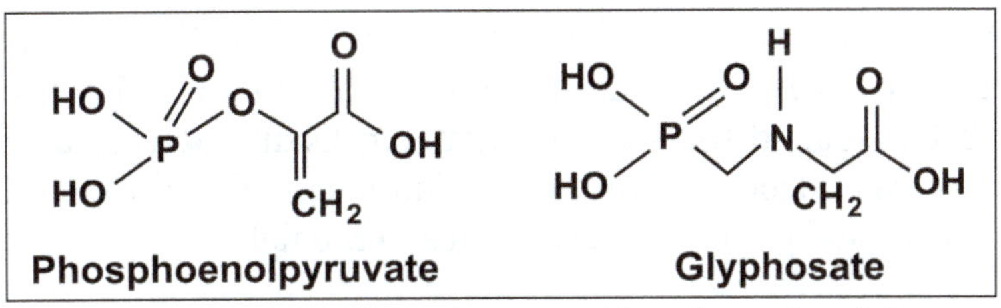

Structures of phosphoenolpyruvate (the substrate) and the herbicide glyphosate (the competitive inhibitor).

This results in inhibition of protein biosynthesis (due to lack of aromatic amino acids). As a consequence, cell division and plant growth are blocked. Further, the plant growth regulator indole acetic acid (an auxin) is also produced from tryptophan. The net result of glyphosate is the death of the plants. Glyphosate is toxic to microorganisms as they also possess shikimate pathway.

Glyphosate is non-toxic to animals (including humans), since they do not possess shikimate pathway. Of the three aromatic amino acids (synthesized in this pathway), tryptophan and phenylalanine are essential and they have to be supplied in the diet, while tyrosine can be formed from phenylalanine.

Strategies for Glyphosate Resistance

There are three distinct strategies to provide glyphosphate resistance to plants:

1. Overexpression of crop plant EPSPS gene: An overexpressing gene of EPSPS was detected in Petunia. This expression was found to be due to gene amplification rather than an increased expression of the gene. EPSPS gene from Petunia was isolated and introduced into other plants. The increased synthesis of EPSPS (by about 40 fold) in transgenic plants provides resistance to glyphosate. These plants can tolerate glyphosate at a dose of 2-4 times higher than that required to kill wild-type plants.

2. Use of mutant EPSPS genes: An EPSPS mutant gene that conferred resistance to glyphosate was first detected in the bacterium Salmonella typhimurium. It was found that a single base substitution (C to 7) resulted in the change of an amino acid from proline to serine in EPSPS. This modified enzyme cannot bind to glyphosate, and thus provides resistance.

The mutant EPSPS gene was introduced into tobacco plants using Agrobacterium Ti plasmid vectors. The transgene produced high quantities of the enzyme EPSPS. However, the transformed tobacco plants provided only marginal resistance to glyphosate. The reason for this was not immediately identified.

It was later known that the shikimate pathway occurs in the chloroplasts while the glyphosate resistant EPSPS was produced only in the cytoplasm. This enzyme was not transported to the chloroplasts, hence the problem to provide resistance. This episode made scientists to realize the importance of chloroplasts in genetic engineering.

In later years, the mutant EPSPS gene was tagged with a chloroplast-specific transit peptide sequence. By this approach, the glyphosate-resistant EPSPS enzyme was directed to freely enter chloroplast and confer resistance against the herbicide.

3. Detoxification of glyphosate: The soil microorganisms possess the enzyme glyphosate oxidase that converts glyphosate to glyoxylate and aminomethylphosponic acid. The gene encoding glyphosate oxidase has been isolated from a soil organism Ochrobactrum anthropi. With suitable modifications, this gene was introduced into crop plants e.g. oilseed rape. The transgenic plants were found to exhibit very good glyphosate resistance in the field.

Use of a Combined Strategy

More efficient resistance of plants against glyphosate can be provided by employing a combined strategy. Thus, resistant (i.e. mutant) EPSPS gene in combination with glyphosate oxidase gene are used. By this approach, there occurs glyphosate resistance (due to mutant EPSPS gene) as well as its detoxification (due to glyphosate oxidase gene).

Phosphinothricin Resistance

Phosphinothricin (or glufosinate) is also a broad spectrum herbicide like glyphosate. Phosphinothricin is more effective against broad-leafed weeds but least effective against perennials. (Phosphinothricin and glufosinate are two names for the same herbicide. However, to avoid confusion between glyphosate and glufosinate, phosphinothricin is more commonly used. Basta Aventis and Liberty are the trade names for phosphinothricin).

Phosphinothricin-a Natural Herbicide

Phosphinothricin is an unusual herbicide, being a derivative of a natural product namely bialaphos. Certain species of Streptomyces produce bialaphos which is a combination of phosphinothricin bound to two alanine residues, forming a tripeptide. By the action of a peptidase, bialaphos is converted to active phosphinothricin.

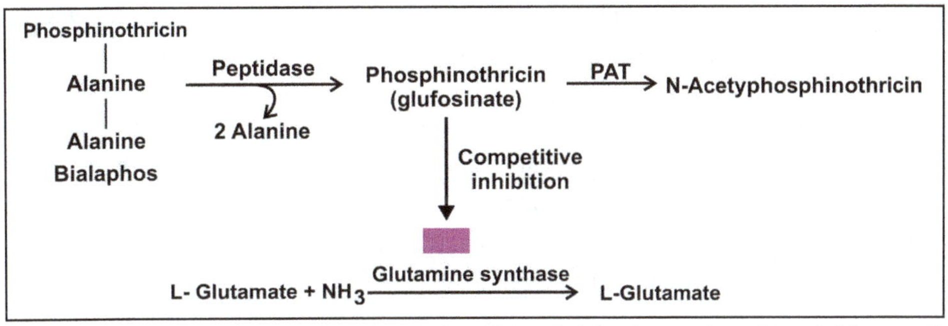

The formation, mode of action and detoxification of phosphinothricin (PAT-Phosphinothrldn acetyl transfetase).

Mechanism of Action of Phosphinothricin

Phosphinothricin acts as a competitive inhibitor of the enzyme glutamine synthase. This is possible since phosphinothricin has some structural similarity with the substrate glutamate. As a consequence of the inhibition of glutamine synthase, ammonia accumulates and kills the plant cells. Further, disturbance in glutamine synthesis also inhibits photosynthesis. Thus, the herbicidal activity of phosphinothricin is due to the combined effects of ammonia toxicity and inhibition of photosynthesis.

Strategy for Phosphinothricin Resistance

The natural detoxifying mechanism of phosphinothricin observed in Streptomyces sp has prompted scientists to develop resistant plants against this herbicide. The enzyme phosphinothricin acetyl transferase (of Streptomyces sp) acetylates phosphinothricin, and thus inactivates the herbicide.

The gene responsible for coding phosphinothricin acetyl transferase (bar gene) has been identified in Streptomyces hygroscopicus. Some success has been reported in developing transgenic maize and oilseed rape by introducing bar gene. These plants were found to provide resistance to phosphinothricin.

Sulfonylureas and Imidazolinones Resistance

The herbicides namely sulfonylureas and imidazolinones inhibit the enzyme acetolactate synthase (ALS), a key enzyme in the synthesis of branched chain amino acids namely isoleucine, leucine and valine. Mutant forms of this enzyme and the corresponding genes have been isolated, identified and characterized. Transgenic plants with the mutant genes of ALS were found to be resistant to sulfonylureas and imidazolinones e.g. maize, tomato, sugar beet.

Resistance to other Herbicides

Some other herbicide resistant plants have also been developed e.g. bromoxynil, atrazine, phenocarboxylic acids, cyanamide. A list of selected examples of gene transferred herbicide resistant plants is given in table.

Table: Selected examples of gene transferred herbicide resistant plants.

Herbicide	Gene transfer/Mechanism of resistance	Transgenic crop(s)
Glyphosate	Inhibition of EPSPS	Soybean, tomato
Glyphosate	Detoxification by glyphosate oxidase	Maize, soybean
Phosphinothricin	bar gene coding phosphinothricin acetyltransferase	Maize, rice, wheat, cotton, potato, tomato, Sugarbeet
Sulfonylureas imidazolinones	Mutant plant with acetolactate synthase	Rice, tomato, maize, sugarbeet
Bromoxynil	Nitrilase detoxification	Cotton, potato, tomato
Atrazine	Mutant plant with chloroplast psb A gene	Soybean
Phenocarboxylic acids	Monooxygenase detoxification (e.g. 2,4-D and 2,4,5-T)	Maize, cotton
Cyanamide	Cyanamide hydratase gene	Tobacco

It may however, be noted that some of the herbicide-resistant transgenic plants are at field-trial stage. Due to environmental concern, a few of these plants are withdrawn e.g. atrazine-resistant crops.

Environmental Impact of Herbicide-Resistant Crops

The development genetically modified (GM) herbicide-resistant crops has undoubtedly contributed to increase in the yield of crops. For this reason, farmers particularly in the developed countries (e.g. USA) have started using these GM crops. Thus, the proportion of herbicide resistant soybean plants grown in USA increased from 17% in 1997 to 68% in 2001.

The farmer is immensely benefited as there is a reduction in the cost of herbicide usage. It is believed that the impact of herbicide-resistant plants on the environment is much lower than the direct use of the herbicides in huge quantities.

There are however, other environmental concerns:

- Disturbance in biodiversity due to elimination of weeds.
- Rapid development of herbicide-resistance weeds that may finally lead to the production of super weeds.

Tolerance to Water Deficit Stresses

The environmental conditions such as temperature (heat, freezing, chilling), water availability (shortage due to drought), and salinity influence the plant growth, development and yield. The abiotic stresses due to temperature, drought and salinity are collectively regarded as water deficit stresses.

Causes of Water Deficit

Water deficit may occur due to the following causes:

- Reduced soil water potential.

- Increased water evaporation (in dry, hot and windy conditions).
- High salt concentration in the soil (decrease soil water potential).
- Low temperature resulting in the formation of ice crystals.

Effects of Water Deficit

- Results in osmotic stress.
- Inhibits photosynthesis.
- Increases the concentration of toxic ions (reactive oxygen species) within the cells.
- Loss of water from the cell causing plasmolysis and finally cell death.

Tolerance to Osmotic Stress

The plant cells are subjected to severe osmotic stress due to water deficit. They however, produce certain compounds, collectively referred to as osmoprotectants or osmolytes, to overcome the osmotic stress. Osmoprotectants are non-toxic compatible solutes and are divided into two groups.

- Sugar and sugar alcohols e.g. mannitol, sorbitol, pinitol, ononitol, trehalose, fructans.
- Zwitterionic compounds: These osmoprotectants carry positive and negative charges e.g. proline, glycine betaine.

The production of a given osmoprotectant is species dependent. The formation of mannitol, proline and glycine betaine are more closely linked to osmotic tolerance.

Strategies to Develop Water Deficit Tolerance Plants

As explained above osmoprotectants offer good protection to plants against osmotic stress and therefore water deficit. It is therefore, logical to think of genetic engineering strategies for the increased production of osmoprotectants.

Some progress has been made in this direction. The biosynthetic pathways for the production of many osmoprotectants have been established and genes coding key enzymes isolated. In fact, some progress has been made in the development of transgenic plants with high production of osmoprotectants.

Transgenic Plants with Glycine toetaine Production

Glycine betaine is a quaternary ammonium compound and is electrically neutral. Besides functioning as a cellular osmolytes, glycine betaine stabilizes proteins and membrane structures. Some of the key enzymes for the production of glycine betaine have been identified e.g. choline mono-oxygenase, choline dehydrogenase, betaine aldehyde dehydrogenase.

The genes coding these enzymes were transferred to develop transgenic plants. By using choline oxidase gene from Arthrobacter sp, transgenic rice that produces higher glycine betaine (which offers tolerance against water deficit stress) has been developed.

Resistance against Ice-Nucleating Bacteria

Formation of ice on the plant cells (outer membrane) is a complex chemical process. The importance of ice-nucleating bacteria is recognized in recent years. The occurrence of these bacteria has been reported in most of the plants – cereals, fruits and vegetable crops. The ice-nucleating bacteria synthesize proteins, which coalesce with water molecules to form ice crystals at temperature around 32°F. As the ice crystals grow, they can pierce the plant cells and severely damage the plants.

Chemical Treatment of Plants to Protect from Ice Formation

Plants can be treated with copper containing compounds to kill the bacteria. Another approach is to use urea solution so that the ice formation is minimized.

Ice-minus Bacteria to Resist Plants from Cold Temperatures

The bacterium Pseudomonas syringae is one of the highly prevalent ice-forming organisms in nature. With genetic manipulations, the gene that directs the synthesis of ice-related bacterial proteins in P. syringae was removed. These newly developed bacteria are referred to as ice-minus bacteria.

The researchers proposed to spray the transgenic ice-minus bacteria on to young plants. The intention was that these bacteria would give frost tolerance to the plants; and thus increase the crop yield. The opponents of DNA technology were against this approach—the main fear being that the bacterial mutants may create some health complication in humans.

The researchers argued and justified that no new genetic information is introduced into P. syringae and it is closely related in all aspect to the parent one which is already in the environment. After prolonged court proceedings in USA, clearance was given for spraying the ice-minus bacteria in the fields.

It was in 1987, ice-minus bacteria were sprayed on to the field of potato plants and strawberry plants. Another strain of P. syringae commercially labeled, as Frostban was later developed and used in crop fields. It may be noted here that ice-minus bacteria of P. syringae were the first transgenic bacteria that were used outside the laboratory. Fortunately, the experiments yielded encouraging results, since crop damage due to frost formation was found to be reduced.

Arabidopsis with Cold-tolerant Genes

Scientists were successful in developing cold- tolerant genes (around 20) in Arabidopsis when this plant was gradually exposed to slowly declining temperatures. They also identified a coordinating gene that encodes a protein, which acts as a transcription factor for regulating the expression of cold-tolerant genes. By introducing the coordinating gene, expression of cold-tolerant genes was triggered, and this protected the plants against cold temperatures. More work is in progress in this direction.

3. Improvement of Crop Yield and Quality

With the advances made in plant genetic engineering, improvement in crop yield and quality have become a reality. The crop yield is primarily dependent on the photosynthetic efficiency and the harvest index (the fraction of the dry matter allocated to the harvested part of the crop). The quality of the crop is dependent on a wide range of desirable characters-nutritional composition of edible parts, flavour, processing quality, shelf-life etc.

Green Revolution

The 'Green Revolution' led by Borlaug, Swaminathan and Khus enabled the world's food supply to be tripled during the last three decades of 20th century. This was made possible by adopting genetically improved varieties of crops, coupled with advances in crop management.

The development of high-yielding varieties of wheat and rice has enabled several developing countries (a good example being India) to move from a position of food scarcity to become net exporter of these cereals.

The Green Revolution became a reality as the farmers adopted to new cereal seeds, besides employing high-input methods of agriculture — use of nitrogen fertilizers, herbicides, pesticides, modern equipment of agriculture etc.

Selected Examples of Crops for Quality and yield

There are a wide range of crops that have been manipulated by scientists for improved yield and quality. Only selected examples are briefly described.

Genetic Engineering for Extended Shelf-Life of Fruits

The genetic manipulation of fruit ripening has become an important commercial aspect in plant genetic engineering.

Delay in Fruit Ripening has many Advantages

- It extends the shelf-life, keeping the quality of the fruit intact.
- Long distance transport becomes easy without damage to fruit.
- Slow ripening improves the flavour.

Genetic engineering work has been extensively carried out in tomatoes, and some of the development are described.

Biochemical Changes during Tomato Ripening

Fruit ripening is an active process. It is characterized by increased respiration accompanied by a rapid increase in ethylene synthesis. As the chlorophyll gets degraded, the green colour of the fruit disappears, and a red pigment, lycopene is synthesized.

The fruit gets softened as a result of the activity of cell wall degrading enzymes namely polygalacturonase (PG) and pectin methyl esterase. The phytohormone ethylene production is intimately linked to fruit ripening as it triggers the ripening process of fruit. Addition of exogenous ethylene promotes fruit ripening, while inhibition of ethylene biosynthesis drastically reduces ripening.

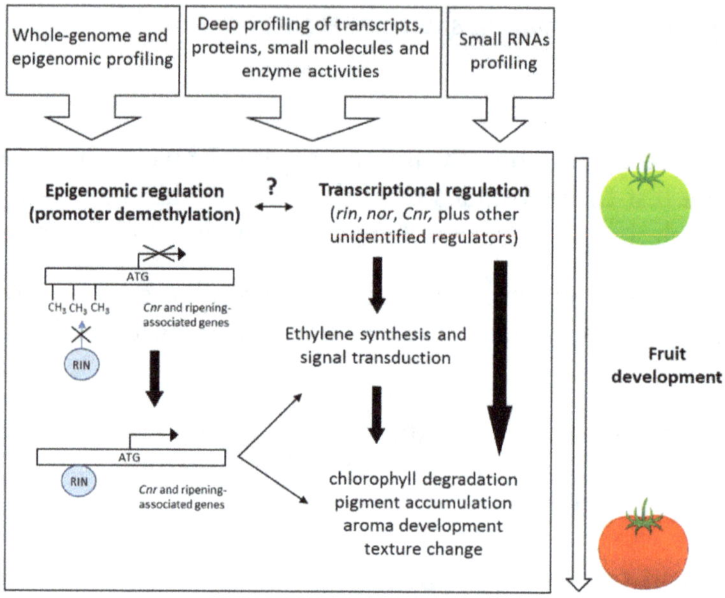

The breakdown of starch to sugars, and accumulation of a large number of secondary products improves the flavour, taste and smell of the fruit. Three distinct genes involved in tomato ripening have been isolated and cloned. The enzymes encoded by these genes and their respective role in fruit ripening are given in table.

Table: Clones of tomato ripening genes and their functions.

Gene clone	Enzyme synthesized	Function in ripening
pTOM 5	Phytoene synthase	Lycopene synthesis that gives red colouration
pTOM 6	Polygalacturonase	Degradation of cell wall, resulting in fruit softening
pTOM 13	ACC oxidase	Ethylene formation that triggers fruit ripening

Genetic Manipulations of Fruit Ripening

Scientists have been trying to genetically manipulate and delay the fruit ripening process. Almost all the attempts involve antisense RNA approach.

Manipulation of the Enzyme Polygalacturonase

As already stated, softening of the fruit is largely due to degradation of the cell wall (pectin) by the enzyme polygalacturonase (PG). The gene responsible for PG, the rotting enzyme, has been cloned (pTOM 6). The genetic manipulation of polygalacturonase by antisense RNA approach for the development of Flavr Savr tomato (by Calgene Company in USA) is depicted in figure and mainly involves the following stages:

- Isolation of the DNA from tomato plant that encodes the enzyme polygalacturonase (PG).
- Transfer of PG gene to a vector bacteria and production of complementary DNA molecules.

- Introduction of complementary DNA into a fresh tomato plant to produce a transgenic plant.

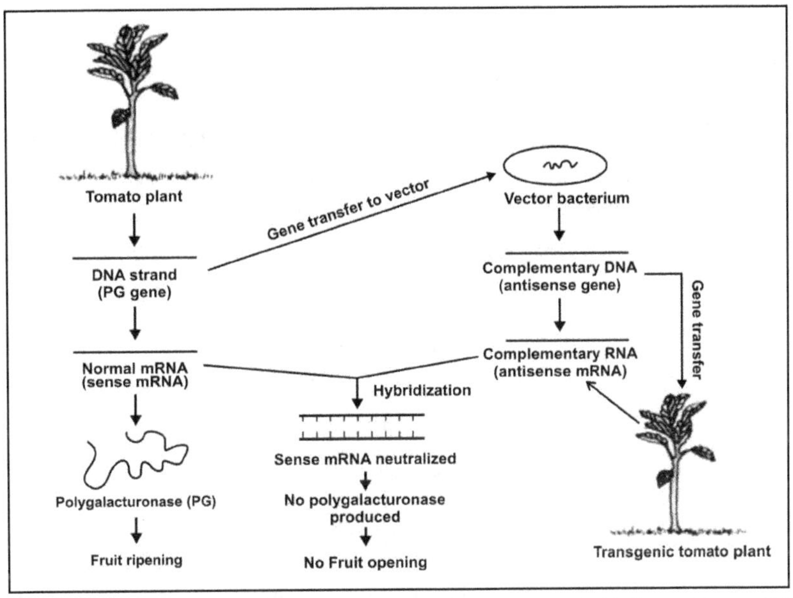

Genetic manipulation of the enzyme potygalacturonase (PG) by antisens RNA approach (Production of Flavr Savr tomato plant).

Mechanism of PG Antisense RNA Approach

In the normal tomato plant, PG gene encodes a normal (sense) mRNA that produces the enzyme polygalacturonase that is actively involved in fruit ripening. The complementary DNA of PG encodes for antisense mRNA, which is complementary to normal (sense) mRNA. The hybridization between the sense and antisense mRNAs renders the sense mRNA ineffective. Consequently, no polygalacturonase is produced, hence fruit ripening is delayed.

The rise and fall of Flavr Savr Tomato

The genetically engineered tomato, known as Flavr Savr (pronounced flavour saver) by employing PC antisense RNA was approved by U.S. Food and Drug Administration on 18th May 1994. The FDA ruled that Flavr Savr tomatoes are as safe as tomatoes that are bred by conventional means, and therefore no special labeling is required. The new tomato could be shipped without refrigeration too far off places, as it was capable of resisting rot for more than three weeks (double the time of a conventional tomato).

Although Flavr Savr was launched with a great fanfare in 1995, it did not fulfill the expectation for the following reasons:

- Transgenic tomatoes could not be grown properly in different parts of U.S.A.
- The yield of tomatoes was low.
- The cost of Flavr Savr was high.

It is argued that the company that developed Flavr Savr, in its overenthusiasm to become the first

Biotech Company to market a bioengineered food had not taken adequate care in developing the transgenic plant. And unfortunately, within a year after its entry, Flavr Savr was withdrawn, and it is now almost forgotten.

Manipulation of Ethylene Biosynthesis

It has been clearly established that ethylene plays a key role in the ripening of fruits. The biosynthetic pathway of ethylene is depicted in figure. Ethylene is synthesized from S-adenosyl methonine via the formation of an intermediate, namely 1 -aminocyclopropane-1 carboxylic acid (ACC), catalysed by the enzyme ACC synthase. The next step is the conversion of ACC to ethylene by ACC oxidase.

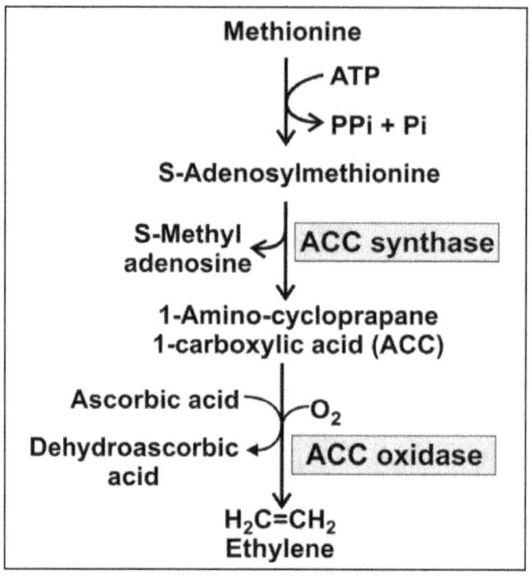

Biosynthesis of ethylene

Three different strategies have been developed to block ethylene biosynthesis, and thereby reduce fruit ripening.

1. Antisense gene of ACC oxidase: Transgenic plants with antisense gene of ACC oxidase have been developed. In these plants, production of ethylene was reduced by about 97% with a significant delay in fruit ripening.

2. Antisense gene of ACC synthase: Ethylene biosynthesis was inhibited to an extent of 99.5% by inserting antisense gene of ACC synthase, and the tomato ripening was markedly delayed.

3. Insertion of ACC deaminase gene: ACC deaminase is a bacterial enzyme. It acts on ACC (removes amino group), and consequently the substrate availability for ethylene biosynthesis is reduced. The bacterial gene encoding ACC deaminase has been transferred and expressed in tomato plants. These transgenic plants inhibited about 90% of ethylene biosynthesis. The fruit ripening was delayed by about six weeks. The strategies 1 and 2 may be referred to as antisense ethylene technology.

Longer Shelf-Life of Fruits and Vegetables

The spoilage of fruits, vegetables and senescence of picked flowers, collectively referred to as post-harvest spoilage is major concern in agriculture. This hampers the distribution system particularly when the transport is done to far off places. The successful manipulations to delay ripening, senescence and spoilage of various foods will significantly contribute to the appropriate food distribution and thus good economic practices in agriculture.

Suppressing the biosynthesis of ethylene appears to be a promising area to reduce the spoilage of fruits, vegetables and senescence of flowers. The three different strategies to block ethylene synthesis in tomato have been described. The same approaches in fact can be successfully used for other fruits, vegetables etc., to achieve longer shelf- life.

Genetic Engineering for Preventing Discoloration

Discoloration of fruits and vegetables is a major postharvest problem encountered in food industry. Certain food additives are added to prevent discoloration. However, these additives may cause health complications in humans. Biochemically, discoloration of fruits and vegetables is mainly due to the oxidation of phenols (mono- and diphenols) to quinones, catalysed by a group of enzymes namely polyphenol oxidases. These enzymes are localized in the membranes of mitochondria and chloroplasts. Genetic manipulations using antisense approach to inhibit the synthesis of polyphenol oxidase has been carried. Some success has been reported in preventing the discoloration of potatoes by this strategy.

Genetic Engineering for Flower Pigmentation

There are continuous attempts in flower industry to make the ornamental flowers more attractive (by improving or creating new colours), besides prolonging post-harvest lifetime. The cut flower industry is mostly (about 70%) dominated by four plants—roses, tulips, chrysanthemums and carnations. The most common type of flower pigments are anthocyanin's, a group of flavonoids. They are synthesized by a series of reactions, starting from the amino acid phenylalanine. The colour of the flower is dependent on the chemical nature of the anthocyanin produced.

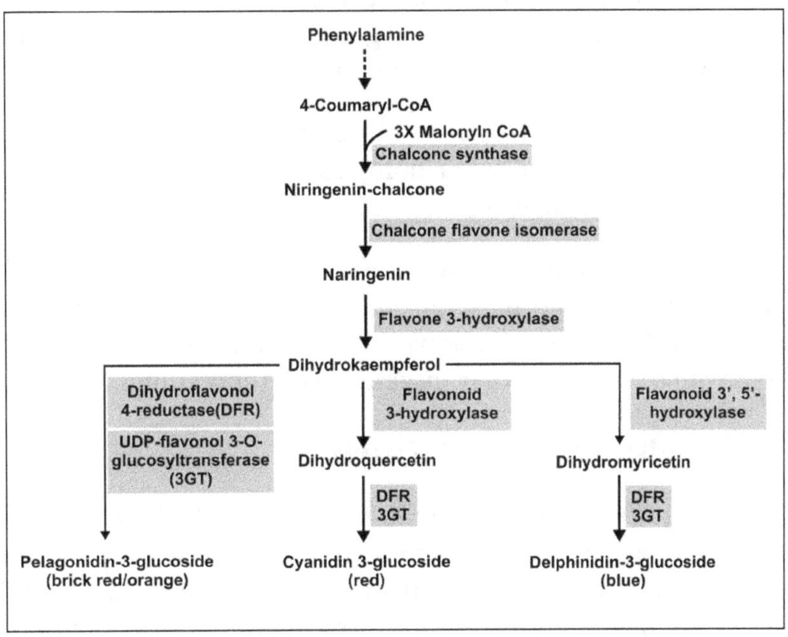

Biosynthesis of anthocyanins.

i. Pelagonidin 3-glucoside — brick red/orange.

ii. Cyanidin 3-glucoside — red.

iii. Delphinidin 3-glucoside — blue to purple.

Manipulation of Anthocyanin Pathway Enzymes

The enzymes responsible for different reactions, in the anthocyanin pathway have been identified. By genetic manipulations and mutations, it is possible to develop flowers with the desired colours. Most of the flowers (roses, carnations chrysanthemums) lack blue colour due to the absence of the key enzyme flavonoids 3′, 5′- hydroxylase (F 3′ 5′ H) that produces delphinidine 3 glucoside. One company, by the name Florigene, has genetically manipulated and introduced the gene encoding the enzyme F 3′ 5′ H (from Petunia hybrida) into the following plants.

The world's first genetically modified (GM) flower was introduced in 1996. It was a mauve (bluish) coloured carnation with a trade name Moondust™. Subsequently, many other flowers have been produced and marketed.

Can GM-flowers be Eaten?

In majority of countries, flowers are used for ornamental purposes and not usually eaten. However, in some countries like Japan, flower petals are used for decoration of foods, and frequently eaten also. This raises an important question about the safety of GM flowers, since they are not thoroughly screened for human consumption. But the present belief is that the anthocyanin's (the colouring chemical molecules) are natural plant materials, and their consumption may be in fact beneficial to health.

Genetic Engineering for Male Sterility

The plants may inherit male sterility either from the nucleus or cytoplasm, cytoplasmic male sterility (cms) is due to the defects in the mitochondrial genome. It is possible to introduce male sterility through genetic manipulations while the female plants maintain fertility.

In tobacco plants, male sterility was introduced by using a mitochondrial mutated gene encoding the enzyme ribonuclease. The gene encoding ribonuclease namely barnase gene from Bacillus amyloliquefaciens was transferred to tobacco plants.

The ribonuclease is toxic to tapetal cells, and thus prevents the development of pollen, ultimately leading to male sterility. By this approach, transgenic plants of tobacco, cauliflower, cotton, tomato, corn, lettuce etc., with male sterility have been developed. It is possible to restore male sterility in the above plants by crossing them with a second set of transgenic plants containing ribonuclease inhibitor gene.

4. Transgenic Plants with Improved Nutrition

Genetic manipulations for improving the nutritional quality of plant products are of great importance in plant biotechnology. Some success has been achieved in this direction through conventional cross-breeding of plants. However, this approach is very slow and difficult, and many a times will not give the traits with the desired improvements in the nutritional quality. Selected examples of genetic engineering with improved nutritional contents are described.

Amino Acids of Seed Storage Proteins

Of the 20 amino acids present in the humans, 10 are essential while the other 10 can be synthesized by the body. The 10 essential amino acids (EAAs) have to be supplied through the diet. Cereals (rice, wheat, maize, corn) are the predominant suppliers of EAAs. However, cereals do not contain adequate quantity of the essential amino acid lysine.

On the other hand, pulses (Bengal gram, red gram, soybean) are rich in lysine and limited in sulfur-containing amino acids (the essential one being methionine). Transgenic routes have been developed to improve the essential amino acid contents in the seed storage proteins of various crop plants.

Overproduction of Lysine by Deregulation

The four essential amino acids namely lysine, methionine, threonine and isoleucine are produced from a non-essential amino acid aspartic acid. The formation of lysine is regulated by feedback inhibition of the enzymes aspartokinase (AK) and dihydrodipicolinate synthase (DHDPS). Theoretically, it is possible to overproduce lysine by abolishing the feedback regulation. This is what has been accomplished.

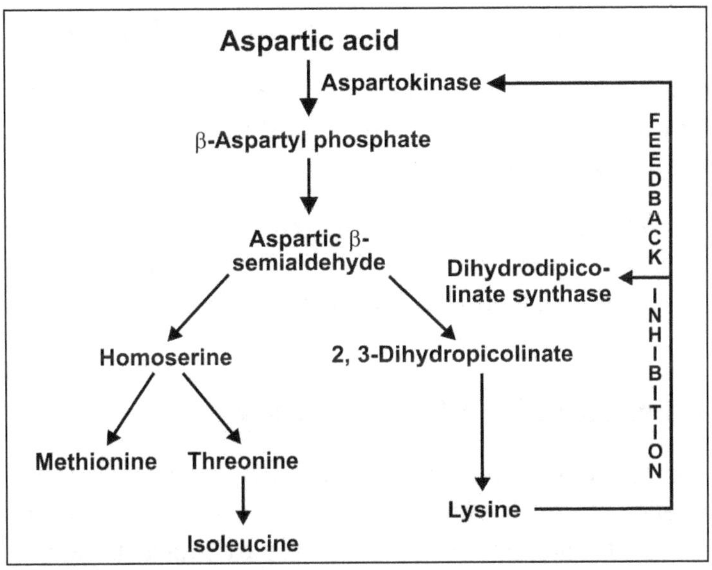

An overview of the biosynthetic pathway of some essential amino acids derived from aspartic acid.

The lysine feedback-insensitive genes encoding the enzymes AK and DHPDS have been respectively isolated from E.coli and Cornynebacterium with appropriate genetic manipulations, these genes were introduced into soybean and canola plants. The transgenic plants so developed produced high quantities of lysine.

Transfer of Genes Encoding Methionine-rich Proteins

Several genes encoding methonine-rich proteins have been identified:

- In maize, 21 KDa zein with 28% methionine.

- In rice, 10 KDa prolamin with 20% methionine.
- In sunflower, seed albumin with 16% methionine.

These genes have been introduced into some crops such as soybean, maize and canola. The transgenic plants produced proteins with high contents of sulfur-containing amino acids.

Production of Lysine-rich Glycinin in Rice

Glycinin is a lysine-rich protein of soybean. The gene encoding glycinin has been introduced into rice and successfully expressed. The transgenic rice plants produced glycinin with high contents of lysine. Another added advantage of glycinin is that its consumption in humans is associated with a reduction in serum cholesterol (hypo- cholesterolaemic effect).

Construction of Artificial Genes to Produce Proteins Rich in EAAs

Attempts are being made to construct artificial genes that code for proteins containing the essential amino acids in the desired proportion. Some success has been reported in the production of one synthetic protein containing 13% methionine residues.

Genetic Engineering for Improving Palatability of Foods

More than the nutritive value, taste of the food is important for attracting humans. It is customary to make food palatable by adding salt, sugar, flavors and many other ingredients. It would be nice if a food has an intrinsically appetizing character.

A protein monellin isolated from an African plant (Dioscorephyllum cumminsii) is about 100,000 sweeter than sucrose on molar basis. Monellin gene has been introduced into tomato and lettuce plants. Some success has been reported in the production of monellin in these plants, improving the palatability.

Golden Rice — The Provitamin A Enriched Rice

About one-third of the world's population is dependent on rice as staple food. The milled rice that is usually consumed is almost deficit in P-carotene, the pro-vitamin A. As such, vitamin A deficiency (causing night blindness) is major nutritional disorder world over, particularly in people subsisting on rice.

To overcome vitamin A deficiency, it was proposed to genetically manipulate rice to produce β-carotene, in the rice endosperm. The presence of β-carotene in the rice gives a characteristic yellow/orange colour, hence the pro-vitamin A-enriched rice are appropriately considered as Golden Rice.

In figure an outline of the biosynthetic pathway for the formation of P-carotene is given. The genetic manipulation to produce Golden Rice required the introduction of three genes encoding the enzymes phytoene synthase, carotene desaturase and lycopene β-cyclase. It took about 7 years to insert three genes for developing Golden Rice.

Golden Rice has met almost all the objections raised by the opponents of GM foods. However, many people are still against the large scale production of Golden Rice, as this will open door to the entry of many other GM foods.

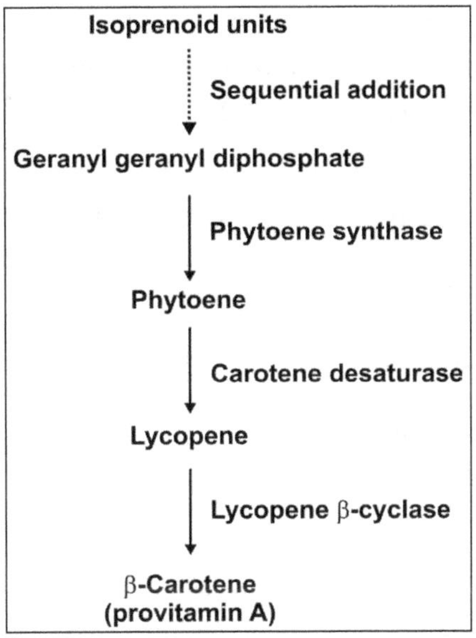

An outline of pathway for the biosynthesis of provitamin A (β-carotene).

Another argument put forth against the consumption of Golden Rice is that it can supply only about 20% of daily requirement of vitamin A. But the proponents justify that since rice is a part of a mixed diet consumed (along with many other foods), the contribution of pro-vitamin A through Golden Rice is quite substantial.

Recently a group of British scientists have developed an improved version of Golden Rice. The new strain, Golden Rice 2 contains more than 20 times the amount of pro-vitamin A than its predecessor. It is claimed that a daily consumption of 70 g rice can meet the recommended dietary allowance for vitamin A.

Genetic Engineering to Increase Vitamins and Minerals

The transgenic rice (Golden Rice) developed with high pro-vitamin A content is described above. Transgenic crop plants are also being developed for increased production of other vitamins and minerals. A transgenic Arabidopsis thaliana that can produce ten-fold higher vitamin E (α-tocopherol) than the native plant has been developed. This was done by a novel approach. A. thaliana possesses the biochemical machinery to produce a compound close in structure to α-tocopherol.

A gene that can finally produce α-tocopherol is also present, but is not expressed. This dormant gene was activated by inserting a regulatory gene from a bacterium. This resulted in an efficient production of vitamin E. Some workers are trying to increase the mineral contents of edible plants by enhancing their ability to absorb from the soil. Some success has been reported with regard to increased concentration of iron.

5. Commercial Transgenic Crop Plants

The very purpose of production of transgenic plants is for their commercial importance with high productivity. It was in 1995-96; transgenic plants (potato and cotton) were, for the first time, made available to farmers in USA. By the year 1998-99, five major transgenic crops (cotton, maize, soybean, canola and potato) were in widespread use. They accounted for about 75% of the total area planted by crops in USA.

A selected list of transgenic crop plants (approved in USA) for the commercial use is given in table. Some examples of transgenic crop plants, which are at the developmental stages are given in table. These plants are carefully designed to give rise to products, which will improve human health and increase of crop yield.

Table: A selected list of transgenic crop plants (GM crops approved in USA) for commercial use.

Crop plant	Genetically altered trait	Product name
Cotton	Insect resistance Glyphosate resistance Bromoxynil resistance Sulfonylurea resistance	Bollguard Roundup ready BXN —
Maize	Insect resistance Insect resistance Glyphosate resistance Glufosinate resistance	Yield Guard Maximizer Roundup Ready Liberty Link
Rice	Vitamin A enrichment	Golden Rice
Tomato	Delayed ripening Delayed ripening Virus resistance	Flavr savr Endless Summer —
Soybean	Glyphosate resistance	Roundup Ready
Potato	Insect resistance Modified starch	Newleaf —
Oilseed rape (canola)	Glufosinate resistance Glyphosate resistance High lauric acid Male sterility hybrid	Innovator Roundup Ready Laurical —
Squash	Virus resistance	Freedom II
Tobacco	Virus resistance	—
Capsicum	Virus resistance	—
Carnation	Modified flower colour	—

Table: Some examples of transgenic crop plants (GM plants) at the developmental stages.

Plant	Gene transfer	Trait transferred/application(s)
For improving human health		
Tomato	Phytoene desaturase	Provitamin A (β-carotene) supplement
Canola	γ- Tocopherol methyl transferase	Vitamin E supplement
Sugar beet	Sucrose-sucrose fructosyl transferase	Fructans-low calorie alternatives to sucrose
Rice	Ferritin	Iron supplement
Potato	Antisense threonine synthase	Increased methionine levels
Potato	Seed albumin	Protein with all essential amino acids
Tomato	S-Adenosylmethionine decarboxylase	Increased lycopene levels
Tomato	Chalcone isomerase	Flavanols-act as antioxidants, reduce risk of cancer, heart diseases
Arabidopsis	Isoflavone synthase	Isoflavones-reduce serum cholesterol, and reduce osteoporosis
Canola	Modified acyl-acyl carrier protein thioesterase	cis-Stearates-lower the risk of heart diseases

For increased crop yield		
Rice	Phosphoenol pyruvate carboxylase	Increased efficiency of photosynthesis
Tobacco	Phytochrome A	Avoids shades
Lettuce	Gibberellic acid (GA) oxidase	Inhibits GA accumulation and stem growth (dwarfing)
Potato	Phytochrome B	Increased photosynthesis and longer life span
Others		
Tobacco and soybean	Cytochrome P_{450}	Synthesis of epoxy fatty acids for manufacture of adhesives and paints
Rice	Nicotianamine aminotransferase	Tolerance to low iron availability
Tobacco	Nitroreductase	Reduces land contamination by trinitrotoluene

Goals of Biotechnological Improvements in Crops

There are about 30-40 crops that have been genetically modified, and many more are being added. However, very few of them have got the clearance for commercial use.

The ultimate goals of genetically modified (CM) crop plants are listed below:

- Resistance to diseases (insect, microorganisms).
- Improved nitrogen fixing ability.
- Higher yielding capacity.
- Resistance to drought and soil salinity.
- Better nutritional properties.
- Improved storage qualities.
- Production of pharmaceutically important compounds.
- Absence of allergens.
- Modified sensory attributes e.g. increased sweetness as in thaumatin.

Concerns about Transgenic Plants

The fears about the harmful environmental and hazardous health effects of transgenic plants still exist, despite the fact that there have been no reports so far in this regard. The transfer of almost all the transgenic plants from the laboratory to the crop fields is invariably associated with legal and regulatory hurdles, besides the social and economic concerns.

The major concern expressed by public (also acknowledged by biotechnologists) is the development of resistance genes in insects, generation of super weeds etc. Several remedial measures are advocated to overcome these problems.

The farmers in developing countries are much worried about the seed terminator technology which forces them to buy seeds for every new crop. These farmers are traditionally habituated to use the seeds from the previous crop which is now not possible due to seed terminator technology.

6. Transgenic Plants as Bioreactors

Another important application of genetically transformed plants is their utility as bioreactors to produce a wide range of metabolic and industrial products.

Characteristics of Genetically Modified Crops

The important characteristics of genetically modified crops are:

Increased Nutrition

Using bioengineering, scientists have added or modified nutrients in various crops, and created several nutritionally enhanced products. Although few have reached commercialization, examples include adding iron to rice, or increasing beta-carotene and vitamin E in vegetable oils to boost the nutritional value. Other genetic modifications have altered the fatty acid composition in oils from soy and canola to create healthier fats.

Plants have also been engineered to increase phytonutrients-substances exclusive of nutrients that have benefits for improving health or preventing disease. These include iso-flavones in soy and lycopene in tomatoes. Two genetic modification strategies have also been devised to increase the iron levels in cereal crops. One is the introduction of the gene that encodes for ferritin, an iron-storage protein. Over expression of this gene improves the storage capacity of plants by as much as three-fold. Using this and other genetic technologies, rice was engineered to contain beta-carotene, which it normally lacks, and enhanced iron content. This transgenic "golden rice" has yet to be bred into hybrid and native strains, so field testing of modified local varieties, commercial production and acceptance are still years away.

Another method for enhancing iron is reducing phytic acid content, which improves the degree and rate at which iron and other minerals are absorbed. In one experiment, corn genetically modified to be low in phytic acid was processed into tortillas. The iron absorption from these tortillas was 49 percent greater than from tortillas made with conventional corn. To further explore the effectiveness of iron absorption by reducing phytic acid, additional iron was added in the form of iron salt supplements and consumed with either strain of corn fed as porridge instead of tortillas. In this case, no absorption effect was observed.

Although it is not clear why the phytic acid level had no effect, it is well known that when dietary iron levels increase, absorption decreases. Other substances in the diet may also have contributed to the reduced absorption. While plants are the primary dietary source of vitamin E, they contain relatively low concentrations of the vitamin. Recent genetic engineering technology has been able to increase the vitamin E content of oils. As it happens, many seeds have abundant levels-up to 20-fold more of gamma-tocopherol, the immediate precursor of alpha-tocopherol, the active form of the vitamin. However, little of the gamma form is converted to the active vitamin.

Researchers identified, isolated and cloned the gene responsible for expressing the enzyme that converts gamma-tocopherol to alpha-tocopherol. The gene was transferred to Arabidopsis, which subsequently exhibited a nine-fold increase in vitamin E. Incorporation of this gene to stimulate similar gamma- tocopherol to alpha-tocopherol conversion into soy, canola and corn is probably not far in the future. Seed oils-particularly mustard and canola-have also been developed to contain carotenoids, especially beta-carotene, a nutrient widely studied for its role in cancer prevention. But this project is still in the testing stage.

Protein (or rather specific amounts of essential amino acids, the building blocks of protein) is needed to fulfil human nutritional requirements for growth, health maintenance and muscle development. In regions of the world where cereal grains cannot be grown, people often rely upon starchy vegetables (roots, tubers or rhizomes) to supply most of their calories. While such crops often have high yields, the primary disadvantage is their very low protein content, less than one percent. Researchers are seeking to improve protein content and quality in vegetable staples such as cassava and plantain through changes in amino acid profiles. For example, a non-allergenic seed albumin gene was introduced into the potato to increase its protein content. Transgenic tubers had 35 to 45 percent more protein and enhanced levels of essential amino acids.

Moreover, transgenic plants produced more tubers and a yield increase of 3 to 3.5 percent. Scientists have also altered soybeans for higher protein in tofu. In an attempt to create healthier fats, researchers have modified the fatty acid composition of soy and canola in several ways. They have produced oils from soy and canola with reduced or zero levels of saturates; canola with medium chain fatty acids; high stearate canola oil free of trans-fatty acids; high oleic acid soybean oil, and canola with the long chain fatty acids gamma linolenic and stearidonic acid.

The latter is of interest as an indirect source of docosahexaenoic acid (DHA), one of two long chain Omega-3 fatty acids shown to be beneficial in protecting against heart attack. DHA is available almost exclusively from seafood, primarily fatty fish. The plant precursor of DHA, linolenic acid, is poorly converted to DHA. Transgenic high oleic acid soybean oil has 80 percent more oleic acid, one-third less saturated fatty acid than olive oil, and no trans-fatty acids. Researchers have also modified sunflower oil for high oleic acid content.

Another type of modified soybean oil is low in saturated fatty acids and richer in linoleic acid than commodity soybean oil. Still another has reduced linolenic acid and no trans-fatty acids, increasing its stability for use as an ingredient in processed foods. Another seed unique for its high level of a single fatty acid is mangosteen. This tropical tree, grown in India, the East Indies and Southeast Asia produces seeds with as much as 56 percent by weight of stearic acid, a saturated fatty acid widespread in foods.

Stearic acid is noteworthy from a nutritional perspective for its stability and textural properties and because it is one of the few saturated fatty acids that does not appear to raise blood cholesterol levels. Thus, it is useful in fats for manufactured and processed foods. Enzymes cloned from mangosteen have also been expressed in canola with resulting increased levels of stearic acid. This research demonstrates the potential of the technology and the unusual sources of enzymes to alter fatty acid profiles in popular food oils such as canola.

Plant biotechnology has also aimed at increasing phytonutrients-substances in plants-exclusive of nutrients that have benefits for improving health or preventing disease. For example, new research in nutrition suggests lutein may support multiple lines of defence against eye disease, and that lycopene serves as a powerful antioxidant in cancer prevention.

Also called "accessory health factors" phytonutrients include iso-flavones in soy, lycopene in tomatoes and polyphenols in green tea. In the laboratory, scientists have engineered tomatoes with 2.5 times as much lycopene as traditional tomatoes. At least one company is developing soy with more iso-flavones, and canola with increased antioxidants and beta-carotenes, lutein and lycopene. There are major constraints on this research, in part because there is still much about phytonutrients that is unknown. For example, some members of a class of phytonutrients may have deleterious effects while others are beneficial, as is the case with various flavonoids, water- soluble plant pigments that, while not considered essential, helps maintain overall health as anti-inflammatory, antihistaminic and antiviral agents.

In addition, scientists do not fully understand the biosynthetic pathways, or the succession of enzyme activities, for many phytonutrients. Another constraint is the limited scientific information about the safety and efficacy of potentially beneficial phytonutrients. Some plants, especially cereals and legumes, are nutritious foods and feeds but also contain varying amounts of substances that interfere with digestibility and nutrient absorption. In excess, these materials may even be toxic.

Genetic modifications are being explored to reduce these anti-nutritional substances, including phytate in cereals and legumes; glycoalkaloids such as solanine and chaconine in potatoes; tomatine, solanine, lectins and oxalate in tomatoes and eggplant; gossypol in cottonseed; trypsin and other protease inhibitors in soy, and tannins and raffinose in legumes. Phytate is widely distributed in cereals and legumes and reduces the absorption of iron, zinc, phosphorus and other minerals in humans and other animals. Phytate is indigestible for swine and poultry because their digestive tracts lack the enzyme phytase, which releases phosphorus from phytate.

Studies have shown that including phytase in the food ration improves phosphorus absorption and reduces phosphorus excretion. In the food animal industry, particularly for swine and poultry, high phytate feeds are associated with high levels of phosphorus excretion. Excess phosphorus in animal manures can be washed into streams or leach into ground water and become a serious source of water pollution. Research has indicated that poultry have substantially reduced phosphorus excretion when fed phytase as a supplement alongside ordinary soybeans or alternatively, genetically transformed soybeans expressing the phytase enzyme. Similarly, swine fed low-phytate corn showed increased phosphorus retention and reduced excretion.

Genetically modified low-phytate corn contains at least five times as much available phosphorus as unmodified corn. Low-phytate corn feed was also associated with improved growth and finishing characteristics. In wheat engineered to expresses the enzyme phytase, seeds exhibited a two to four-fold increase in phytase activity. This opens the possibility of improving the digestibility of wheat, especially among non-ruminant animals.

Scientists are also seeking ways to reduce toxic substances such as glycoalkaloids. Researchers inserted antisense genes into potatoes to block the activity of the enzyme UDP-glucose glucosyl transferase, key to the production of the glycoalkaloid alpha-chaconine. This toxic substance can, at high enough levels, cause irritation of the gastrointestinal tract or impairment of the nervous system. Preliminary findings indicated that the transgenic potatoes produced fewer glycoalkaloids.

Allergens Modification

Food allergies and sensitivities cause a wide variety of conditions, symptoms and diseases, a few of which can be life threatening. A food allergy or hypersensitivity is one that provokes an immune response, while a food intolerance incites an abnormal physiological reaction. Experts estimate that 2 percent of adults, and from 2 to 8 percent of children, are truly allergic to certain foods. Food intolerance is a much more common problem than allergy.

Unlike allergies, intolerances generally intensify with age. The eight most commonly allergenic foods are milk, eggs, peanuts, soybeans, fish, crustaceans, tree nuts and wheat. There are also significant allergies to non-food plants, such as ryegrass and other plants with airborne pollens that may cause hay fever or other seasonal allergic symptoms. Most known allergens in food are proteins, suggesting the possibility of modifying the structure, or possibly eliminating the allergenic protein from the food. In some cases, traditional plant breeding has identified hypoallergenic strains that are targets for further genetic modification to reduce allergenicity.

Neutralizing the allergens in major food grains would have an enormous impact on millions of families, where one or more members cannot eat these foods that are household staples. Researchers have used this approach in rice, the first food crop with reduced allergenicity to be created through genetic engineering. Further testing and development work continues to assure that people with known allergies to rice products can consume this genetically engineered food with-out developing their typical allergic reaction. In foods such as peanuts, however, which are highly allergenic to some sensitive individuals, the allergenic proteins constitute the majority of the plant's protein, so that elimination may not be possible.

Another example where genetic modification may be used to reduce allergenicity is in wheat, one of the "big eight" allergenic foods. Although not yet commercially available, scientists have genetically engineered wheat to overexpress the gene responsible for the synthesis of thioredoxin, an enzyme that catalyses the reduction of disulfide bonds within protein molecules, thus reducing the protein's allergenic properties. When expressed in wheat, the enzyme reduced the bonds in the major allergenic proteins the gliadens and glutenins and to a lesser extent the minor ones, too, making them markedly less allergenic. At the same time, the functional characteristics of the wheat were not impaired.

Scientists are also exploring the potential of recombinant DNA technology to reduce the allergenicity of non-food allergens. For example, ryegrass is a dominant source of airborne pollen in temperate climates, and using antisense technology, scientists engineered ryegrass with reduced Lol p 5 protein levels. As this is the major allergen in ryegrass, the modification reduced the plant's allergenicity. Although genetic engineering has the potential to reduce allergenicity of foods, it also has the potential for unintentionally introducing new allergens.

Improved Functional Properties

Researchers are in search of enhanced functional properties for specific purposes, such as firmer tomatoes for canning, or beans with less breakage. One of the first applications to reach the market was the highly publicised Flavr Savr tomato, which was genetically engineered for delayed ripening. While the transformation process did delay ripening and extend shelf life, the product was expensive to produce and purchase and some consumers did not like the taste. This led to its withdrawal from the market. Other work aims to create a tomato that ripens on the vine but remains firm during harvest, handling and shipping.

Firm tomatoes are preferable for canning, which consumes the largest share of tomato production. Using antisense technology, researchers have created tomatoes that are 40 percent firmer than their conventional counterparts and stay firm for at least two weeks. Scientists have also engineered beans for desirable canning characteristics such as firm texture and seed coats that do not split. Several experiments are being conducted with soybeans. One would diminish that undesirable byproduct of bean consumption, flatulence, by creating high sucrose soybeans through reduction of the carbohydrate raffinose.

Another seeks to modify soybean oil to reduce the linoleic acid content so that it is more stable for industrial applications. Presently, barley is an unsuitable feed for poultry because poultry lack the enzyme to break down β-glucan, the predominant polysaccharide (a type of carbohydrate) in endosperm cell wails. Scientists have created transgenic malt that can depolymerise β-glucan. Adding transgenic malt to barley-based poultry feed enabled poultry to metabolize barely, grow as well as poultry fed a corn-soybean diet, and produce more hygienic droppings. The digestibility of feeds can also be improved with modification of starch levels in different crops. For example, cattle can more readily digest amylose- free wheat in feed.

Extensive research has been directed toward altering the properties, quantity and distribution of starch in many plants, for a variety of purposes. The principal forms of starch are either linear (amylose) or branched polymers amylopectin. Using genetic engineering technology to influence the amount and length of chain branching and polymerisation increases the availability of starches with different properties. It also enables the development of novel starches.

However, plants differ widely in where they store different types of starch; thus modification of starch production must be tailored to the particular plant. What works in the potato, for example, may not work in wheat or rice. Moreover, results in one variety of a crop may not be obtained in another. A well-known example of the modulation of starch synthesis has been the development of transgenic potatoes engineered to contain a gene for an enzyme affecting starch synthesis. The transgenic potatoes had up to 60 percent more starch than non-engineered strains. The increased starch content made the potatoes take up less fat during frying, resulting in a lower-fat product.

About 40 percent of tapioca starch is used for the production of modified starch, sweeteners and the flavour enhancer monosodium glutamate. In processing tapioca, a significant amount of starch remains in the waste material and wastewater. It is estimated that even after extraction, the waste still contains 50 percent starch. Bioengineered improvements in tapioca, such as reduction in water content and higher starch concentration, may increase the ease of processing the plant material into a finished starch product.

Further, raising the efficiency of starch utilisation in the processing of sweeteners reduces the amount of starch reaching the waste stream. The possibility of converting the starch content of wastewater to energy, using high rate anaerobic digestion, is promising.

However, a number of factors remain to be overcome, including the effect of environmental sulfates in the waste stream and the efficiency of energy production. The use of transgenic organisms offers potential solutions. When the enzyme thioredoxin is overexpressed in barley endosperm, the activity of the enzyme pullulanase, a rate-limiting enzyme in breaking down starch, increases four-fold. Breaking down starch is a key part of the barley malting process, and tests with this engineered variety showed that the time required could be reduced by up to a day. Over expression of thioredoxin also hastened barley germination, of special interest to growers of this normally slow-germinating grain.

Permissions

All chapters in this book are published with permission under the Creative Commons Attribution Share Alike License or equivalent. Every chapter published in this book has been scrutinized by our experts. Their significance has been extensively debated. The topics covered herein carry significant information for a comprehensive understanding. They may even be implemented as practical applications or may be referred to as a beginning point for further studies.

We would like to thank the editorial team for lending their expertise to make the book truly unique. They have played a crucial role in the development of this book. Without their invaluable contributions this book wouldn't have been possible. They have made vital efforts to compile up to date information on the varied aspects of this subject to make this book a valuable addition to the collection of many professionals and students.

This book was conceptualized with the vision of imparting up-to-date and integrated information in this field. To ensure the same, a matchless editorial board was set up. Every individual on the board went through rigorous rounds of assessment to prove their worth. After which they invested a large part of their time researching and compiling the most relevant data for our readers.

The editorial board has been involved in producing this book since its inception. They have spent rigorous hours researching and exploring the diverse topics which have resulted in the successful publishing of this book. They have passed on their knowledge of decades through this book. To expedite this challenging task, the publisher supported the team at every step. A small team of assistant editors was also appointed to further simplify the editing procedure and attain best results for the readers.

Apart from the editorial board, the designing team has also invested a significant amount of their time in understanding the subject and creating the most relevant covers. They scrutinized every image to scout for the most suitable representation of the subject and create an appropriate cover for the book.

The publishing team has been an ardent support to the editorial, designing and production team. Their endless efforts to recruit the best for this project, has resulted in the accomplishment of this book. They are a veteran in the field of academics and their pool of knowledge is as vast as their experience in printing. Their expertise and guidance has proved useful at every step. Their uncompromising quality standards have made this book an exceptional effort. Their encouragement from time to time has been an inspiration for everyone.

The publisher and the editorial board hope that this book will prove to be a valuable piece of knowledge for students, practitioners and scholars across the globe.

Index

Q

R

S

T

V

www.ingramcontent.com/pod-product-compliance
Lightning Source LLC
LaVergne TN
LVHW080704210926
848485LV00003B/415